SpringerBriefs in Applied Sciences and Technology

Computational Mechanics

For further volumes:
http://www.springer.com/series/8886

Severino P. C. Marques · Guillermo J. Creus

Computational Viscoelasticity

 Springer

Severino P. C. Marques
Centro de Tecnologia
Universidade Federal de Alagoas
Maceió, Alagoas
Brazil
e-mail: smarques@lccv.ufal.br

Guillermo J. Creus
ILEA
Universidade Federal do Rio Grande do Sul
Porto Alegre
Rio Grande do Sul
Brazil
e-mail: creus@ufrgs.br

ISSN 2191-5342
ISBN 978-3-642-25310-2
DOI 10.1007/978-3-642-25311-9
Springer Heidelberg Dordrecht London New York

e-ISSN 2191-5350
e-ISBN 978-3-642-25311-9

Library of Congress Control Number: 2011943323

Printed on acid-free paper

Springer is part of Springer Science+Business Media (www.springer.com)

To my wife, Dilze, and my children,
Fernando, Gustavo and Clarissa

S. P. C. Marques

To my wife, Susana, and my children,
Guillermo, Tomás, Amalia and Ana

G. J. Creus

Preface

This book develops a presentation of viscoelasticity theory oriented toward numerical applications. It is our hope that it will be useful both as a textbook for graduate courses and as a reference volume for engineers and researchers.The book is structured in twelve chapters. The first eight chapters introduce basic concepts and theoretical ideas about the viscoelastic response of solids. They cover constitutive relations in integral and differential form, influence of temperature, age and finite strain. These topics were selected aiming to make the access to the computational viscoelastic formulations easier. It is assumed that the reader has a background in mathematics and mechanics at the undergraduate level. In the last five chapters a more advanced experience may be needed.

The remaining chapters address the numerical formulation of viscoelastic problems using finite element, boundary element and finite volume methods. Chapter 9 presents viscoelastic finite element procedures formulated on a total Lagrangian description for large displacements and rotations with small strains. Two alternative boundary element procedures for the solution of problems in linear viscoelasticity are reviewed in Chap. 10: the solution in the Laplace transformed domain and the use of a general inelastic formulation. Chapter 11 presents a two-dimensional approach for linear viscoelastic solids using a finite volume framework. Together with the theoretical formulations, worked examples are presented throughout the text. Finally, in Chap. 12, further examples, to be solved with the software Abaqus, are proposed and developed. The book concludes with three Appendices which contain auxiliary expressions in mathematics and mechanics.

Several colleagues and students provided essential help. We mention here professors L. A. B. Cunda (FURG), B. F. Oliveira (UFRGS) and Paul Partridge (UnB). D. La Porta, D. Palmer and R. Sprunger (SIMULIA) helped with the Abaqus examples, Litha Bacci draw the figures and Joice de Brito e Cunha checked the English text. This work is the result of collaboration between the Federal University of Alagoas (UFAL) and the Federal University of Rio Grande do Sul (UFRGS) with the financial support of the Brazilian Agency CAPES through PROCAD program. The continuous support of our research by the Brazilian Agency CNPq is also gratefully acknowledged.

Federal University of Alagoas-Brazil   Severino P. C. Marques
Federal University of Rio Grande do Sul-Brazil   Guillermo J. Creus

Contents

Chapter 1
Introduction

1.1 Historical Context

First studies. It took time to discover that the properties of important materials lay outside the classical limits of Hookean elastic solids and Newtonian viscous fluids. Tests on the mechanical properties of silk threads, performed in 1835 by Wilhelm Weber, showed that solid behavior could have viscous components. Later, in 1867, James Clerk Maxwell introduced elastic properties in the description of fluids. Boltzmann developed in 1874 the formulation for linear viscoelasticity. Using the superposition of effects, he showed that the strain at time t in response to a general time-dependent stress history $\sigma(t)$ can be written as the sum (or integral) of terms that involve the strain response to a step loading. The mathematician Vito Volterra [7] developed the theory of functional and integral equations adequate to model viscoelastic phenomena. Differential and integral representations of viscoelasticity [2, 3, 6] are addressed in Chap. 2.

Further developments and problems. The developments in the first half of the twentieth century were slow and important advances in theoretical and experimental rheology took place only after World War II. New materials, such as polymers and composite materials [1, 5] posed new problems, particularly the need to solve boundary value problems in varying conditions of temperature and humidity. Chapter 3 introduces the state variables formalism, important for efficient computation and Chap. 4 extends the viscoelastic formulation to three dimensional situations. The effect of temperature is studied in Chap. 6, and the Laplace transform technique, used to solve boundary value problems, is reviewed in Chap. 5. In the analysis of materials such as rubber, soft polymers and biological tissues strains are large and it is necessary to dispense with the infinitesimal strain theory. To maintain objectivity in the presence of large rotations, measures like the Cauchy-Green tensor for strain and the Piola–Kirchhoff tensors for stress are introduced. This formulation is reviewed in Chap. 8. Biological tissues [4],

S. P. C. Marques and G. J. Creus, *Computational Viscoelasticity*,
SpringerBriefs in Computational Mechanics, DOI: 10.1007/978-3-642-25311-9_1,
© The Author(s) 2012

polymers and other important materials show a mechanical behavior that depends on age. This subject is introduced in Chap. 7.

Computational viscoelasticity. Digital computers revolutionized the practice of many areas of engineering and science, and solid mechanics was among the first fields to use them. Many computational techniques have been used in this field, but the one that emerged in the 1970s as the most widely adopted is the Finite Element Method. This method was developed and put to practical use for the analysis of aeronautical structures by Ray W. Clough and J. H. Argyris. In the most common version of the Finite-Element Method, the domain to be analyzed is divided into elements, and the displacement field within each element is interpolated in terms of the displacements at the nodes. From the displacements, strains and stresses are calculated in terms of nodal displacements. The equilibrium equations expressed through the principle of virtual work generate a system of simultaneous equations to be solved by the computer. With the Finite Elements Method, for the first time, real problems could be analyzed considering the actual geometry and material properties. First bar structures and small strain elasticity and then geometrical and physical nonlinear problems were addressed and solved. Lately, both the Boundary Element Method, that reduces the dimension of the problems and provides very precise results, and the Finite Volume Method, which seems to be very efficient for the study of non-homogeneous solids, were developed. These numerical procedures are analyzed in the second part of this book, Chaps. 9, 10, 11. In Chap. 12 some computational examples and exercises are included, using Abaqus software.

1.2 Basic Experimental Results

The characteristic feature of viscoelastic behavior is the essential role played by time. Viscoelastic materials under constant stress increase their deformation with time, while, under constant strain, show stresses that decrease with time. Figure 1.1 indicates the behavior of a typical viscoelastic material in a *creep test* characterized by the application of a constant stress σ_0 at a time τ_0.

Using the unit step function $H(t)$, defined in Appendix A, we may write this stress history as

$$\sigma(t) = \sigma_0 H(t - \tau_0) \tag{1.1}$$

which defines both the value of the applied stress and the time of its application. In a creep test we measure an elastic strain component ε^e(instantaneous) and a creep (delayed) component ε^c. The latter is the one that increases with time and characterizes viscoelastic behavior. The deformation that remains after $\tau > \tau_0$ characterizes hysteresis.

Removing the applied stress at time $\tau_1 > \tau_0$, that is, considering the stress history

$$\sigma(t) = \sigma_0 H(t - \tau_0) - \sigma_0 H(t - \tau_1) \tag{1.2}$$

Fig. 1.1 Creep test of a viscoelastic solid: histories of stress and strain. Full line, loading; dotted line, unloading

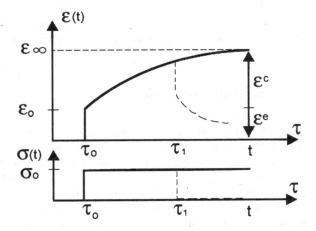

Fig. 1.2 Relaxation test of a viscoelastic solid: histories of stress and strain. Full line: loading; dotted line: unloading

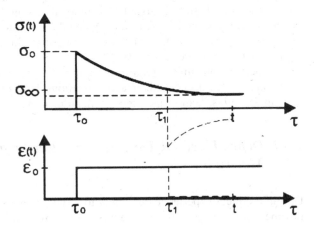

we obtain for $t > \tau_1$ the strain history shown by the dotted line in Fig. 1.1. The deformation reduction upon unloading is known as *creep recovery*.

In a *relaxation test* we have the material subjected to an imposed constant deformation

$$\varepsilon(t) = \varepsilon_0 H(t - \tau_0) \qquad (1.3)$$

and we measure the stress $\sigma(t)$ that is needed to keep strain at the constant value ε_0. We observe that $\sigma(t)$ diminishes progressively, as indicated by the stress history in Fig. 1.2. Removing the applied deformation at time τ_1, that is, considering the history

$$\varepsilon(t) = \varepsilon_0 H(t - \tau_0) - \varepsilon_0 H(t - \tau_1) \qquad (1.4)$$

we obtain the stress history shown in dotted lines. It is interesting to observe that in this case we may have a change in the sign of the resultant stress; this fact may be of importance for materials with different strengths in tension and compression.

1.3 Constitutive Relations

The general principles of mechanics (e.g. equilibrium and compatibility equations, thermodynamic principles) are valid for all materials. The characteristic properties of each material are specified by its constitutive equations.

A *constitutive equation* is a relation between forces and deformations. In popular terms, the forces applied to a body "cause" it to deform and the quality and amount of deformation varies according to the nature of the body. In the present context (small deformation analysis) stresses and deformations are conveniently represented by *Cauchy stress* σ and *infinitesimal strain* ε. Constitutive relations will be firstly discussed in a uniaxial setting. The extension to the multiaxial case will be analyzed in Chap. 4, and the extension to finite strains, in Chap. 8. A more precise definition of the concepts of strain and stress can be found in Appendix B and references there.

In practice, constitutive relations are firstly suggested by experiments and then established by means of mathematical equations. New experiments, new materials, new applications, lead to new more refined or more sophisticated models.

1.3.1 Dependence on Time History; Elastic and Viscoelastic Materials

During a typical experience, we apply to a specimen a stress history $\sigma(t)$, variable in time ($\tau_0 \leq t \leq \infty$) and we measure the corresponding strain history $\varepsilon(t)$. We may also apply a deformation history $\varepsilon(t)$ and measure the resulting stresses $\sigma(t)$, because the choice of the controlled variable is a matter of experimental convenience. For an arbitrary stress history, the strain at time t will depend, in general, upon all the values of stress in the time interval of the experiment, so that we can write

$$\varepsilon(t) = \mathcal{D}\{\sigma(\tau)\}_{\tau=\tau_0}^{\tau=t} \tag{1.5}$$

where $\mathcal{D}$ indicates a functional $\mathcal{D} : C(\tau_0, t) \Rightarrow R$ while $C(\tau_o, t)$ and R indicate respectively the set of continuous functions defined in the interval $[\tau_0, t]$ and the set of real numbers. Eq. (1.5) indicates that the value of ε at time t depends on all the values of $\sigma(\tau)$ for τ varying between τ_0 and t. τ_0 is an arbitrary initial time, so that $\sigma(t) = 0$ and $\varepsilon(t) = 0$ for $t < \tau_0$.

Similarly, we can write

$$\sigma(t) = \mathcal{E}\{\varepsilon(\tau)\}_{\tau=\tau_0}^{\tau=t} \tag{1.6}$$

Notice here the existence of two symbols representing time. t is used to represent the time of interest. For example, in (1.5) we are interested in the deformation at time t. This deformation depends on all the stresses applied to the material in different instants up to time t. To avoid confusion, we use another symbol, τ, to represent those instants. τ is a dummy variable that runs in the interval that ends in t.

A different functional $\mathcal{D}$ corresponds to each class of material. For example, in elastic materials the deformation at time t depends on the value of the stress at the same time t: instantaneous, non hereditary response. Elastic materials have a very short memory: they recall only the present stress, when $\tau = t$. In this case, the functional in (1.5) is reduced to an ordinary function and

$$\varepsilon(t) = D(\sigma(t)). \tag{1.7}$$

If the material is linearly elastic (1.7) may be still simplified to

$$\varepsilon(t) = D\sigma(t) \tag{1.8}$$

where D is now a constant factor, the *elastic compliance*, which is the inverse of the *elastic modulus E*.

On the other hand, viscoelastic materials are characterized by a dependence on the whole history of the deformation process, and their constitutive relations must have the functional structure indicated in (1.5) and (1.6). Considering for example the creep test, as described in Sect. 1.2, we see that its result may be expressed in the form (1.5). In this particular case, the argument of the functional is completely determined once we know the values of σ_0, t and τ, being τ a generic time for loading. Thus, creep tests may be characterized by a functional whose argument is formed by step functions, or, equivalently, by a function of three variables

$$\varepsilon(t) = D(\sigma_0, t, \tau_0) \tag{1.9}$$

We have already seen how this function depends on t; now we will analyze its dependence on σ_0(the stress applied in the creep test) and τ_0(the time at which the creep test begins).

1.3.2 Dependence on Stress: Linearity

Figure 1.3a indicates the stress and strain histories for creep tests of a typical material at different stress levels. We see that for small stresses the deformations tend to stabilize, while for high stresses they grow at an increasing rate. This type of behavior is usual in concrete, polymers and many other viscoelastic materials. Figure 1.3b shows isochronous curves, that are obtained from Fig. 1.3a by setting

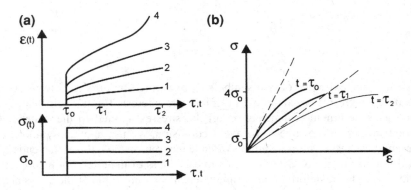

Fig. 1.3 a Creep tests with different values of stress and the corresponding strain histories.
b isochronous curves corresponding to the tests in (a)

τ as a parameter. We can do this graphically just by choosing values of τ in
Fig. 1.3a and determining the corresponding values of σ and ε. These *isochronous*
(from iso: equal, chronos: time) curves are pseudo stress–strain relations, but of
course are valid only in reference to creep tests.

In the case of Fig. 1.3, the threshold of nonlinearity is about $2\sigma_0$. Its precise
location depends on the accepted tolerance. Linearity in this context may be
characterized by superposition. Consider arbitrary stress histories of the type

$$\sigma(\tau) = \sigma_1(\tau) + \sigma_2(\tau) \,; \; \tau \in [\tau_0, t] \qquad (1.10)$$

If they give rise to strain histories that can be expressed as

$$\varepsilon(\tau) = \varepsilon_1(\tau) + \varepsilon_2(\tau) \qquad (1.11)$$

where $\varepsilon_1(\tau)$ and $\varepsilon_2(\tau)$ are the strain histories corresponding to $\sigma_1(\tau)$ and $\sigma_2(\tau)$
separately, we say that the material is linear. Linear behavior is also referred to as
obeying the "*Principle of superposition in viscoelasticity*" or "*Boltzmann principle*". To check linear behavior experimentally, step functions are usually used.
The representation of nonlinear viscoelasticity is addressed in Chap. 8.

On the linear range, we may write (1.9) in the form

$$\varepsilon(t) = \sigma_0 D(t, \tau) \qquad (1.12)$$

where $D(t, \tau)$, *the specific creep function* or *creep compliance*, defined as the
response at time t to a unit step of stress applied at time τ, fully characterizes the
behavior of a linear viscoelastic material.

In material testing it is usual to use uniaxial tension or compression loading
applying a strain history with constant rate $\varepsilon(t) = vt$. An elastic material will show
a stress history also with constant rate. This is not the case when the material is
linear viscoelastic. A typical result is shown in Fig. 1.4. Stress-time and stress–
strain relations are not linear except for very slow ($v \to 0$) or very fast ($v \to \infty$)
loading rates.

Fig. 1.4 Loading of a linear
viscoelastic material
(standard model) with
constant strain rate

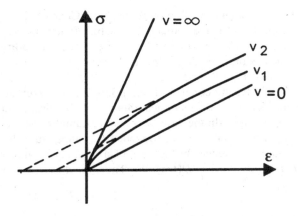

Fig. 1.5 Surface
representing the creep
function $D(t, \tau)$ for a material
that hardens with age (i.e.
concrete)

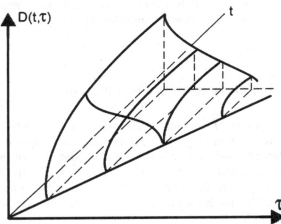

When the limiting value of $D(t, \tau)$ is finite, i.e., $\lim_{t\to\infty} D(t, \tau) = M(\tau) < \infty$ we say
that the material is asymptotically stable. Sometimes, asymptotically stable
materials are referred to as solids, while those materials for which $D(t, \tau)$ grows
indefinitely are called fluids. For stable materials we have

$$\frac{\partial D(t, \tau)}{\partial (t - \tau)} < 0 \tag{1.13}$$

1.3.3 Dependence on Age: Aging

We call *aging* the change in the mechanical properties of a given material due to
its age, where age is the time period between some origin more or less arbitrarily

established and the time of observation. Concrete is a typical example of an aging material. From the moment of casting (taken usually as age zero) it begins to increase its strength and to decrease its deformability. The function $D(t, \tau)$ that indicates the specific creep has for concrete the form indicated in Fig. 1.5. Notice that $D(t, \tau) = 0$ for $t < \tau$.

Frequently, the concept of aging involves other influences in addition to elapsed time. Aging is different according to the environmental conditions in which the material ages. In the case of concrete, humidity and temperature are important. In the case of polymers factors such as temperature, humidity, UV radiation, etc., make a difference. In the case of a viscoelastic material without aging we have $D(t + a, \tau + a) = D(t, \tau)$, $\forall a$. Thus, for $a = -\tau$, we can write

$$D(t, \tau) = D(t - \tau) \tag{1.14}$$

Non-aging materials represent a special (very important) case of viscoelastic materials. Additional formulations and examples for the *aging* case are given in Chap. 7.

1.4 State Variables Formulation

Besides the functional representation described in Sect. 1.3, a state variables representation may be used with some advantages in viscoelasticity as well as in plasticity and damage mechanics. An advantage of the *state variable approach* is that physical theories, and micro-structural information, may be introduced directly in the formulation of the evolution equations. Another one is that it leads to more efficient numerical procedures. This formulation will be introduced in Chap. 3.

1.5 Computational Viscoelasticity

Because of mathematical difficulties few real problems in viscoelasticity have analytical solution. As in many other areas of science, the use of numerical analyses and digital computers had a great impact in this field. Procedures based on techniques as Finite Elements, and more recently, Boundary Elements and Finite Volumes allow the analysis of complex bodies and structures made of linear and nonlinear viscoelastic materials. In Chaps. 9, 10, 11 of this book, these numerical procedures are described. To allow the reader to have some practice with computational procedures a few examples using the well known commercial software Abaqus are given.

References

1. E.J. Barbero, *Finite Element Analysis of Composite Materials* (CRC Press, Boca Raton, 2008)
2. R.M. Christensen, *Theory of Viscoelasticity* (2nd edn. Dover Publications, New York, 2010)
3. W. Flugge, *Viscoelasticity* (Springer-Verlag, New York, 1975)
4. Y.C. Fung, *Biomechanics: mechanical Properties of Living Tissues* (Springer Science +Business Media, LLC, New York, 2004).
5. R.S. Lakes, *Viscoelastic Solids* (CRC Press LLC, Boca Raton, 1999)
6. N. Rabotnov Yu, *Elements of hereditary solid mechanics* (MIR Publishers, Moscow, 1980)
7. V. Volterra, *Theory of functionals* (Dover, New York, 1959)

Chapter 2
Rheological Models: Integral and Differential Representations

Viscoelastic relations may be expressed in both integral and differential forms. Integral forms are very general and appropriate for theoretical work. Differential forms are related to rheological models that provide a more direct physical interpretation of viscoelastic behavior. In this chapter we describe the most usual rheological models, deduce their differential equations and, by solving them, we find the corresponding integral representations. These relations will be set in a more computational friendly form in Chap. 3 and extended to three-dimensional situations in Chap. 4 and then used in analytical and computational solutions.

2.1 General Integral Relations

When the functional relation (1.5) in Chap. 1 is linear it has a simple and useful representation given by the Riesz theorem [1]: if the functional D is *linear* and *equi-continuous*, it can be written

$$\varepsilon(t) = \int_{\tau_0}^{t} D(t - \tau)d\sigma(\tau) \tag{2.1}$$

or

$$\varepsilon(t) = \int_{\tau_0}^{t} D(t - \tau)\dot{\sigma}(\tau)d\tau \tag{2.2}$$

Here, τ_0 should be chosen in a way that for $\tau < \tau_0$ the material is at rest, without stress and strain. From the relations above we see that $D(t - \tau)H(t - \tau_0)$ represents

S. P. C. Marques and G. J. Creus, *Computational Viscoelasticity*,
SpringerBriefs in Computational Mechanics, DOI: 10.1007/978-3-642-25311-9_2,
© The Author(s) 2012

the strain corresponding to a creep test with $\sigma(t) = H(t - \tau_0)$. $D(t - \tau)$ is the creep function or creep compliance of dimension L^2/F.

Equation (2.1) is an integral of the Stieltjes type. For these integrals, when $\sigma(\tau)$ has steps $\Delta\sigma_i H(t - \tau_i)$ we have

$$\varepsilon(t) = \int_{\tau_0}^{t} D(t - \tau)d\sigma(\tau) + \sum_i \Delta\sigma_i D(t - \tau_i) \qquad (2.3)$$

As long as $\sigma(t)$ is continuous and differentiable, $\dot{\sigma}(t)$ exists and the form (2.2) can be used. Notice that the integration is performed with relation to τ; t acts as a parameter and as the superior limit of integration, but it is a constant inside the integral. Thus, for differentiation in relation to t we have to use the Leibnitz formula (See Appendix A).

Alternative forms of the integral representation. Besides the relations (2.1) and (2.2) we may use the inverse relations

$$\sigma(t) = \int_{\tau_0}^{t} E(t - \tau)d\varepsilon(\tau) \qquad (2.4)$$

and

$$\sigma(t) = \int_{\tau_0}^{t} E(t - \tau)\dot{\varepsilon}(\tau)d\tau \qquad (2.5)$$

exchanging the roles of stress and strain. $E(t - \tau)$ is the specific relaxation function, i.e., the stress response to a unit step of strain. Integrating (2.2) by parts, we obtain

$$\varepsilon(t) = \frac{\sigma(t)}{E(0)} + \int_{\tau_0}^{t} d(t - \tau)\sigma(\tau)d\tau \qquad (2.6)$$

where

$$d(t - \tau) = -\frac{\partial}{\partial\tau}D(t - \tau)$$
$$E(0) = 1/D(t - t) \qquad (2.7)$$

Sometimes, instantaneous and delayed components of the specific creep are separated

$$D(t - \tau) = \frac{1}{E(0)} + C(t - \tau)$$
$$E(t - \tau) = E(\infty) + R(t - \tau) \qquad (2.8)$$

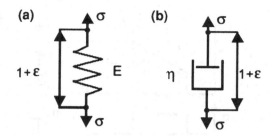

Fig. 2.1 a Hooke model
(*spring*). **b** Newton model
(*dashpot*)

2.2 Rheological Models

The behavior of viscoelastic materials under uniaxial loading may be represented by means of conceptual models composed of elastic and viscous elements which provide physical insight and have didactic value. Rheological models are described in most of the books on viscoelasticity such as Flugge [2], Christensen [3] and many others.

2.2.1 The Basic Elements: Spring and Dashpot

An ideal helicoidal *spring*, perfectly linear *elastic* and massless, represents Hooke model (see Fig. 2.1a):

$$\sigma(t) = E\varepsilon(t) \tag{2.9}$$

where E is the elasticity modulus with dimension $[F/L^2]$. Both length and cross-section are given unit values in order to identify force with stress and elongation with strain.

The *dashpot* (Fig. 2.1b) is an ideal *viscous* element that extends at a rate proportional to the applied stress, according to Newton equation

$$\dot{\varepsilon}(t) = \sigma(t)/\eta \tag{2.10}$$

where $\dot{\varepsilon} = \partial\varepsilon/\partial t$ is the *rate of strain* and η is the *viscosity coefficient*, with dimension $[FT/L^2]$. Combining springs and dashpots we obtain different models of viscoelastic behavior. The simplest viscoelastic models are those named after the scientists J. C. Maxwell and Lord Kelvin.

2.2.2 Maxwell Model

This model is the combination of a spring and a dashpot in series, Fig. 2.2a. For this system we may write the equations

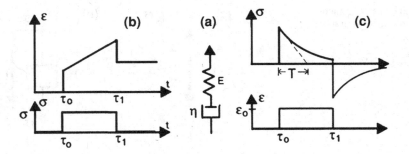

Fig. 2.2 Maxwell model: **a** rheological model, **b** creep test, **c** relaxation test

$$\varepsilon(t) = \varepsilon_E(t) + \varepsilon_\eta(t)$$
$$\sigma_E(t) = \sigma_\eta(t) = \sigma(t) \tag{2.11}$$
$$\sigma_E(t) = E\varepsilon_E(t); \sigma_\eta(t) = \eta\dot{\varepsilon}_\eta(t)$$

where the sub-indexes η and E indicate dashpot and spring respectively.

Differentiating the first Eq. 2.11 with respect to time t and using the constitutive relations for both spring and dashpot, we obtain

$$\dot{\varepsilon}(t) = \frac{\dot{\sigma}(t)}{E} + \frac{\sigma(t)}{\eta} \quad \varepsilon(t) = \sigma(t) = 0 \text{ for } t < \tau_0 \tag{2.12}$$

which is the differential equation for the Maxwell model. Solutions of (2.12) may be determined considering either stress or strain as the controlled variable. In the first case we have directly

$$\varepsilon(t) = \frac{\sigma(t)}{E} + \frac{1}{\eta} \int_{\tau_0}^{t} \sigma(\tau)d\tau \tag{2.13}$$

Integrating (2.13) by parts we obtain the alternative expression

$$\varepsilon(t) = \int_{\tau_0}^{t} \left(\frac{1}{E} + \frac{t-\tau}{\eta} \right) \dot{\sigma}(\tau)d\tau \tag{2.14}$$

Comparing this to (2.2) we see that

$$D(t-\tau) = \frac{1}{E} + \frac{t-\tau}{\eta}; \quad t \geq \tau \tag{2.15}$$

is the *creep function*. Since the strain response is unbounded for $t \rightarrow \infty$, one says that the Maxwell model exhibits unbounded creep and sometimes refers to it as Maxwell *fluid*. For a stress history $\sigma(t) = \sigma_0[H(t-\tau_0) - H(t-\tau_1)]$, with

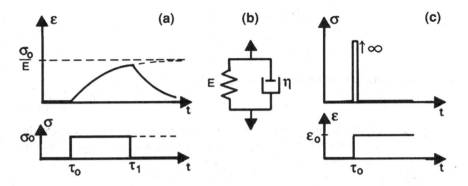

Fig. 2.3 Kelvin model: **a** creep test, **b** rheological model, **c** relaxation test

$\tau_0 < \tau_1$, as that shown in Fig. 2.2b, a residual deformation remains after unloading.

Considering now the strain history as given, we obtain from (2.12), using the general solution for first order differential equations in Appendix A,

$$\sigma(t) = E \int_{\tau_0}^{t} e^{-\frac{E}{\eta}(t-\tau)}\dot{\varepsilon}(\tau)d\tau \qquad (2.16)$$

Then, comparing to (2.5) we see that

$$E(t - \tau_0) = Ee^{-(t-\tau_0)/T}; \ T = \frac{\eta}{E}; \ t \geq \tau_0 \qquad (2.17)$$

is the relaxation function that vanishes for $t \to \infty$. Relations using creep or relaxation functions, such as (2.14) and (2.16), are equivalent. A procedure to obtain one from the other is given in Sect. 2.4 and in Chap. 5.

The constant $T = \eta/E$ that appears in the exponential in (2.17) determines the rate of the relaxation process and is called *relaxation time*. The smaller the relaxation time, the faster the relaxation process, even though total relaxation takes theoretically an infinite time. For example, for $t - \tau_0 = 3T$ about 95% of the total relaxation is completed. Considering a loading–unloading history, such as $\varepsilon(t) = \varepsilon_0[H(t - \tau_0) - H(t - \tau_1)]$, with $\tau_0 < \tau_1$, the stress response changes signal (Fig. 2.2c).

2.2.3 Kelvin Model

This model combines a spring and a dashpot in parallel, Fig. 2.3b. From the relations

$$\sigma(t) = \sigma_E(t) + \sigma_\eta(t)$$
$$\varepsilon_E(t) = \varepsilon_\eta(t) = \varepsilon(t) \tag{2.18}$$
$$\sigma_E(t) = E\varepsilon_E(t); \sigma_\eta(t) = \eta\dot{\varepsilon}(t)$$

we can determine the differential equation

$$\sigma(t) = E\varepsilon(t) + \eta\dot{\varepsilon}(t) \tag{2.19}$$

For a given strain history we have the stress directly from (2.19). A relaxation test is physically impossible with the Kelvin model because $\dot{\varepsilon}(t) = \varepsilon_0\delta(t)$ and the corresponding initial stress should be infinitely high.

For a given stress history $\sigma(t)$ the solution of (2.19) is

$$\varepsilon(t) = \frac{1}{\eta}\int_{\tau_0}^{t} \sigma(\tau)e^{-\frac{t-\tau}{\theta}}d\tau \quad ; \quad \theta = \frac{\eta}{E} \tag{2.20}$$

Comparing to (2.2) we see that

$$D(t - \tau_0) = \frac{1}{E}\left(1 - e^{-(t-\tau_0)/\theta}\right); \quad t \geq \tau_0 \tag{2.21}$$

is the creep function for the Kelvin model. For $t \to \infty$ we obtain $\varepsilon(\infty) = \sigma_0/E$ that corresponds to the *asymptotic elastic solution*, when all the stress is carried by the spring.

Again, we have equivalent differential and integral representations. Fig. 2.3 shows the results of creep and relaxation tests. The constant θ is called *retardation time* and is analogous in meaning to the relaxation time: an estimate of the time required for the creep process to approach completion.

2.3 Generalized Models

Maxwell and Kelvin models are adequate for qualitative and conceptual analyses, but generally poor for the quantitative representation of the behavior of real materials. In order to improve the representation we need to increase the number of parameters by combining a number of springs and dashpots. A systematic way to do that is to build generalized Maxwell and Kelvin models, shown in Fig. 2.4. The *generalized Maxwell model* is composed of $n + 1$ constituent elements in parallel, being n Maxwell models and an isolated spring (to warrant solid behavior) (see Fig. 2.4a).

The differential Eq. 2.12 for a generic Maxwell element r of a generalized Maxwell model may be written in the operational form

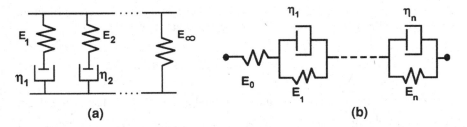

Fig. 2.4 Maxwell and Kelvin chains with instantaneous elasticity

$$\frac{\partial}{\partial t}\varepsilon(t) = \left(\frac{1}{E_r}\frac{\partial}{\partial t} + \frac{1}{\eta_r}\right)\sigma_r \qquad (2.22)$$

where E_r, η_r and σ_r indicate the elastic constant, viscosity coefficients and stress of the *r-th* element, respectively. The symbol $\partial/\partial t$ is a differential operator that can be handled as an algebraic entity. For the generalized Maxwell model the strain is the same for all constituent elements and the total stress is given by the equation

$$\sigma(t) = \left(E_\infty + \sum_{r=1}^{n}\frac{\partial/\partial t}{\frac{\partial/\partial t}{E_r} + \frac{1}{\eta_r}}\right)\varepsilon(t) \qquad (2.23)$$

From Fig. 2.4 and (2.17) it is clear that the relaxation function for the generalized Maxwell model is, for a generic value of τ

$$E(t - \tau) = E_\infty + \sum_{r=1}^{n} E_r e^{-\frac{t-\tau}{T_r}} \quad ; \quad T_r = \eta_r/E_r \qquad (2.24)$$

The generalized Maxwell model provides an exponentially varying stress adding contributions with different relaxation times, one for each element in the chain. Thus, it is possible to fit experimental creep curves to any required degree of approximation if enough terms are used. To find the creep function for the generalized Maxwell model the differential Eq. 2.23 has to be solved, like in Example 1 below.

The *generalized Kelvin model* is composed of n Kelvin units in series plus an isolated spring. The stress at each unit is the same external stress $\sigma(t)$ while the total (observable) strain $\varepsilon(t)$ is the sum of the internal strains in each element. Writing (2.19) in the symbolic form for a generic Kelvin element r

$$\sigma_r(t) = \left(E_r + \eta_r{\partial}\!/{\partial t}\right)\varepsilon_r(t) \qquad (2.25)$$

we have for the model in Fig. 2.4b

$$\varepsilon(t) = \left(\frac{1}{E_0} + \sum_{r=1}^{n}\frac{1}{E_r + \eta_r\partial/\partial t}\right)\sigma(t) \qquad (2.26)$$

From Eq. 2.21 and Fig. 2.4b, it is easy to gather that the specific creep function for the generalized Kelvin model is, for a generic value of τ,

$$D(t - \tau) = \frac{1}{E_0} + \sum_{r=1}^{n} \frac{1}{E_r}[1 - e^{-\frac{t-\tau}{\theta_r}}] \quad ; \quad \theta_r = \eta_r/E_r \tag{2.27}$$

To find the relaxation function, the differential equation (2.26) has to be solved.

Example 1 Determine the differential equation of the *Zener model*, that is a particular case of the generalized Maxwell model composed by a Maxwell model with parameters $E_1 = E$, $\eta_1 = \eta$ in parallel with a spring of stiffness E_∞, Fig. 2.4a. Substituting these values into (2.23) we obtain

$$\sigma(t) = \left(E_\infty + \frac{\partial/\partial t}{\frac{\partial/\partial t}{E} + \frac{1}{\eta}} \right) \varepsilon(t) \tag{2.28}$$

Developing this symbolic equation we find

$$\sigma + \frac{\eta}{E}\dot{\sigma} = E_\infty\varepsilon + \frac{\eta(E_\infty + E)}{E}\dot{\varepsilon} \tag{2.29}$$

With $E_{z0} = E_\infty + E$, $\theta_z = \eta(E_\infty + E)/(E_\infty E) = \eta E_{z0}/[E_\infty(E_{z0} - E_\infty)]$ and $T_z = \eta/E = \eta/(E_{z0} - E_\infty)$, we have the nice form

$$E_z(0)\left[\dot{\varepsilon}(t) + \frac{\varepsilon(t)}{\theta_z} \right] = \dot{\sigma}(t) + \frac{\sigma(t)}{T_z} \tag{2.30}$$

where $E_z(0) = E_{z0}$. Solving in ε we obtain, with the initial condition $\varepsilon(\tau_0) = \sigma(\tau_0)/E_{z0}$,

$$\varepsilon(t) = \frac{\sigma(t)}{E_z(\infty)} - \left[\frac{1}{E_z(\infty)} - \frac{1}{E_z(0)} \right] \int_0^t e^{-\frac{(t-\tau)}{\theta_z}}\dot{\sigma}(\tau)d\tau \tag{2.31}$$

being $E_z(\infty) = E_\infty$. The corresponding creep function is then

$$D(t - \tau) = \frac{1}{E_z(\infty)}\left[1 - \frac{E_z(0) - E_z(\infty)}{E_z(0)}e^{-\frac{(t-\tau)}{\theta_z}} \right] \tag{2.32}$$

Example 2 Determine the differential equation of the *standard solid* model which is a particular case of the generalized Kelvin model with a spring (E_0) and a Kelvin element ($E_1 = E$, $\eta_1 = \eta$) connected in series. Substituting these parameters in (2.26), we have

$$\varepsilon(t) = \left(\frac{1}{E_0} + \frac{1}{E + \eta\partial/\partial t} \right)\sigma(t) \tag{2.33}$$

Developing this equation, the following differential equation is obtained

$$\sigma + \frac{\eta}{E_0 + E}\dot{\sigma} = \frac{E_0 E}{E_0 + E}\varepsilon + \frac{E_0 \eta}{E_0 + E}\dot{\varepsilon} \qquad (2.34)$$

Making $E_s(0) = E_0$, $E_s(\infty) = E_0 E/(E_0 + E)$, $\theta_s = \eta/E$ and $T_s = \eta/(E_0 + E)$, this differential equation can be written as

$$E_s(0)\left[\dot{\varepsilon}(t) + \frac{\varepsilon(t)}{\theta_s}\right] = \dot{\sigma}(t) + \frac{\sigma(t)}{T_s} \qquad (2.35)$$

Comparing (2.35) to (2.30), we conclude that the standard and Zener models present similar differential equations. Then, the solution for each one of these models can be obtained from the solution of the other by a convenient change of parameters.

2.3.1 General Differential Representation

Equations 2.23 and 2.26 are differential equations with the general form

$$\sum_{i=0}^{h} p_i \frac{\partial^i \sigma}{\partial t^i} = \sum_{j=0}^{k} q_j \frac{\partial^j \varepsilon}{\partial t^j} \qquad (2.36)$$

where p_i and q_j are material constants dependent on the viscoelastic model. Usually, without loss of generality, we assume $p_0 = 1$.

From (2.30), the constants for the Zener model are

$$p_0 = 1, p_1 = \frac{\eta}{E}, q_0 = E_\infty \text{ and } q_1 = \frac{\eta(E_\infty + E)}{E} \qquad (2.37)$$

and, for the standard solid model (2.35),

$$p_0 = 1, p_1 = \frac{\eta}{E_0 + E}, q_0 = \frac{E_0 E}{E_0 + E} \text{ and } q_1 = \frac{E_0 \eta}{E_0 + E} \qquad (2.38)$$

Generalized Kelvin and Maxwell models are equivalent, in the sense that it is always possible to find a generalized Maxwell model equivalent to a given generalized Kelvin one, as in Examples 1 and 2. In Chap. 5 it is shown how to go from a creep to a relaxation function. Then, with Eqs. (2.24) and (2.27) we can find the corresponding models.

2.4 Integral and Differential Operators

Viscoelastic relationships may also be indicated in the symbolic forms

$$\varepsilon = D^* \sigma$$
$$\sigma = E^* \varepsilon \tag{2.39}$$

which have to be interpreted as alternatives to (2.1) and (2.4) respectively. As the linear operators E^* and D^* may be handled formally as algebraic quantities (with some care), this notation simplifies some calculations. The operational form is valid also for the differential representation. For example, the differential operator for the generalized Kelvin model is the expression inside the brackets in (2.26). With this notation viscoelastic and elastic equations have similar form. A more rigorous development and applications of the operational technique will be given in Chap. 5 through the use of *Laplace transform*.

Sometimes we need to invert the viscoelastic relations, i.e., to obtain the relaxation function corresponding to a given creep function and vice versa.

From (2.39), we have

$$\varepsilon = D^* E^* \varepsilon \tag{2.40}$$

Thus,

$$H(t - \tau_0) = D^* E(t - \tau) H(t - \tau_0) \tag{2.41}$$

In extended form, this is written [4, 5]

$$1 = D(t - \tau)E(t - t) + \int_{\tau_0}^{t} D(t - \tau)\dot{E}(\tau - \tau_0)d\tau \quad \text{for } t \geq \tau_0 \tag{2.42}$$

Equation (2.42) express the obvious fact that applying as a stress history the corresponding relaxation function, we obtain a constant unit deformation.

Example 3 Consider the relaxation function corresponding to the Zener model with $E(t) = E_1 + E_2 e^{-t/T}$; then $\dot{E}(t) = -E_2 e^{-t/T}/T$ and substituting into (2.42)

$$D(t - \tau_0)(E_1 + E_2) - \frac{E_2}{T} \int_{\tau_0}^{t} D(t-\tau)e^{-(t-\tau)/T}d\tau = 1 \tag{2.43}$$

Differentiating (2.43) in relation to t (Leibnitz rule)

$$\dot{D}(t - \tau_0)(E_1 + E_2) - \frac{E_2}{T}D(t - \tau_0) + \frac{E_2}{T^2} \int_{\tau_0}^{t} D(t-\tau)e^{-(t-\tau)/T}d\tau = 0 \tag{2.44}$$

Multiplying this equation by T and adding to (2.43) we eliminate the integral to obtain the differential equation

$$(E_1 + E_2)T\dot{D}(t - \tau_0) + E_1 D(t - \tau_0) = 1 \tag{2.45}$$

From which we obtain, with the initial condition $D(0) = 1/E(0) = 1/(E_1 + E_2)$

$$D(t - \tau_0) = \frac{1}{E_1} - \frac{E_2}{(E_1 + E_2)E_1} e^{-(t-\tau)/\theta} \tag{2.46}$$

with $\theta = (1 + E_2/E_1)T$.

The operational form is valid also for the differential representation. For example, the differential operator for the generalized Kelvin model is the expression inside the brackets in (2.26).

2.5 Thermodynamic Restrictions

The work done in deforming a viscoelastic body must be non-negative. Sufficient conditions that the relaxation function must satisfy are given in [6]. In reference to Eq. (2.5)

(1) $E(t)$ must be non-negative
(2) $E(t)$ must be a monotonically decreasing function with finite limit for $t \to \infty$.
(3) $E(t)$ must be convex downward.

Most of the functions that are usually used to approximate the relaxation function satisfy the conditions above.

References

1. F. Riesz, B. Sz.-Nagy, *Functional Analysis* (Dover Publications Inc., New York, 1990)
2. W. Flugge, *Viscoelasticity* (Springer, New York, 1975)
3. R.M. Christensen, *Theory of Viscoelasticity*, 2nd edn. (Dover Publications Inc, New York, 2010)
4. G.J. Creus, *Viscoelasticity-Basic Theory and Application to Concrete Structures* (Springer, Berlin, 1986)
5. A.C. Pipkin, *Lectures on Viscoelasticity Theory* (Springer, Heidelberg, 1972)
6. S. Breuer, E. Onat, On uniqueness in linear viscoelasticity. Q. Appl. Math. **19**(4), pp 355–359 (1962)

Chapter 3
State Variables Approach

The State Variables approach is an alternative to the history dependent integral representation given in Chap. 2. It has a physical basis because of its origin in thermodynamic formulations and shows computational advantages. The creep and relaxation functions are approximated by exponential series. The introduction of state variables leads to n differential equations of first order in place of the differential equation of order n linked to a generalized model. This formulation leads to exponential expressions that make incremental integration easier, allowing the determination of the viscoelastic strains at time $t + \Delta t$ as a function of the viscoelastic strains and stresses at time t. Then, there is no need to store the whole history of stress or strain. In this chapter, we introduce the basic formulation that is later extended to 3D, aging and nonlinear situations.

3.1 Basic Formulation

We have already seen (Chap. 2) that for each rheological model, we may have an integral as well as a differential relation. We shall see now how to obtain, through the State Variables approach, a new general form convenient for computational applications.

In some important commercial computer codes (i.e. Abaqus, Ansys) the theoretical formulation is introduced in the integral form. Then, reference is made to a rheological model (generalized Maxwell) and Prony series are introduced as its representation. The State Variables approach [1, 3] presented here helps us to understand better the relation among these formulations.

Let us begin with the general form of the integral representation (see (2.6))

S. P. C. Marques and G. J. Creus, *Computational Viscoelasticity*,
SpringerBriefs in Computational Mechanics, DOI: 10.1007/978-3-642-25311-9_3,
© The Author(s) 2012

$$\varepsilon(t) = \frac{\sigma(t)}{E_0} + \int_{\tau_0}^{t} d(t - \tau)\sigma(\tau)d\tau \tag{3.1}$$

where $E_0 = E(\tau_0)$.

We may approximate the function $d(t - \tau)$ by means of an exponential series (called Dirichlet-Prony series by mathematicians)

$$d(t - \tau) = \sum_{i=1}^{n} d_i e^{-\frac{(t-\tau)}{\theta_i}} \tag{3.2}$$

Such an approximation is *complete* (i.e. can be as good as we like, depending on the number of terms included). We introduce then the n values

$$q_i(t) = \int_{\tau_0}^{t} d_i e^{-\frac{(t-\tau)}{\theta_i}} \sigma(\tau)d\tau \quad i = 1, \ldots, n \tag{3.3}$$

and notice that, by taking the differential of (3.3) with relation to t (using Leibnitz rule, Appendix A) we obtain

$$\dot{q}_i(t) + \frac{q_i(t)}{\theta_i} = d_i\sigma(t) \quad i = 1, \ldots, n \tag{3.4}$$

which is a system of n *uncoupled linear differential equations of the first degree* that, with the adequate initial conditions (e.g. $q_i = 0$ for $t = \tau_0$) allows us to determine the state variables. Then, with

$$\varepsilon(t) = \frac{\sigma(t)}{E_0} + \sum_{i=1}^{n} q_i(t) \tag{3.5}$$

that results from (3.1), (3.2) and (3.3) the strain can be determined.

Moreover, we notice that:

(i) Equations (3.4) and (3.5) correspond in this case to a generalized Kelvin Model with springs of constants $1/d_i\theta_i$ and dashpots with constants $1/d_i$ and an isolated spring of constant E_0.

(ii) From (3.4) and (3.5) we see also that, when the values q_i are known at a given instant t_r, then $\varepsilon(t)$ may be determined for $t \geq t_r$ if $\sigma(t)$ is known for $t \geq t_r$. Thus, if the strains are measured for $t \geq t_r$ with respect to the configuration at t_r, the material behavior in the interval (t_r, t_f) with $t_f > t_r$ will depend only on the value of the stresses in (t_r, t_f) and the value of q_i at time t_r. The observations above justify the name *state variables* representation.

(iii) This representation is very general and can be extended to aging and non-linear problems as well (see Chaps. 7 and 8).

(iv) State variables may also be related to the material microstructure [5].

(v) Finally, this formulation is convenient for the numerical solution of viscoelastic problems (using for example the Finite Element Method).

Alternatively, we may begin with the relation

$$\sigma(t) = \int_{\tau_0}^{t} E(t - \tau)\dot{\varepsilon}(\tau)d\tau \tag{3.6}$$

and expand

$$E(t) = E_\infty + \sum_{i=1}^{n} E_i e^{-t/T_i} \tag{3.7}$$

Introducing the n quantities

$$q_i(t) = \int_{\tau_0}^{t} e^{-\frac{(t-\tau)}{T_i}} \dot{\varepsilon}(\tau)d\tau \quad i = 1, \ldots, n \tag{3.8}$$

and taking the derivative of (3.8) with respect to t we obtain

$$\dot{q}_i(t) + \frac{q_i(t)}{T_i} = \dot{\varepsilon}(t) \quad i = 1, \ldots, n \tag{3.9}$$

A combination of (3.6), (3.7) and (3.8) provides the equation

$$\sigma(t) = E_\infty \varepsilon(t) + \sum_{i=1}^{n} E_i q_i(t) \tag{3.10}$$

Equations (3.9) and (3.10) together with the initial conditions $q_i = 0$ when
$t = \tau_0$ correspond to a Maxwell generalized model with springs E_i, dashpots $T_i E_i$
and an isolated spring E_∞. Here q_i represents the strains in the springs of the
Maxwell elements.

3.2 Incremental Determination of State Variables

We may integrate system (3.4) in a variety of ways.

1. The use of the simple Euler process gives us, for a step increment Δt, the
algorithm

$$\Delta q_i(t) = -\left(\frac{q_i(t)}{\theta_i} - d_i\sigma(t)\right)\Delta t$$
$$q_i(t + \Delta t) = q_i(t) + \Delta q_i(t) \tag{3.11}$$

2. Better results are obtained writing

$$q_i(t + \Delta t) = \int_{\tau_0}^{t} d_i e^{-(t+\Delta t - \tau)/\theta_i} \sigma(\tau) d\tau + \int_{t}^{t+\Delta t} d_i e^{-(t+\Delta t - \tau)/\theta_i} \sigma(\tau) d\tau \qquad (3.12)$$

The first integral is equal to $q_i \exp(-\Delta t/\theta_i)$; the second one may be written, assuming $\sigma(t)$ constant over the interval $[t, t + \Delta t]$, as

$$d_i \sigma(t) e^{-(t+\Delta t)/\theta_i} \int_{t}^{t+\Delta t} e^{\tau/\theta_i} d\tau = d_i \theta_i \sigma(t)(1 - e^{-\Delta t/\theta_i}) \qquad (3.13)$$

so that

$$q_i(t + \Delta t) = e^{-\Delta t/\theta_i} q_i(t) + d_i \theta_i \sigma(t)(1 - e^{-\Delta t/\theta_i}) \qquad (3.14)$$

Expressions of this type are used in some viscoelastic computer codes (e.g. Ansys). Simo and Hughes [6] propose an alternative of the same accuracy. The value of a state variable at a given time may be determined directly from its value at the previous time step and the history of stress or strain; this incremental procedure is more accurate than the Euler integration procedure and is more efficient than the direct numerical calculation of the integral in (3.3).

3. A more efficient algorithm may be obtained considering the relation $\sigma(\tau)$ to vary linearly in the interval $[t, t + \Delta t]$

$$\sigma(\tau) = \sigma(t) + \overline{\Delta\sigma}(t)(\tau - t); \quad t \le \tau \le t + \Delta t \qquad (3.15)$$

where

$$\overline{\Delta\sigma}(t) = [\sigma(t) - \sigma(t - \overline{\Delta t})]/\overline{\Delta t} \qquad (3.16)$$

and $\overline{\Delta t}$ is the time interval in the preceding step. Then, (3.16) takes the form

$$q_i(t + \Delta t) = e^{-\Delta t/\theta_i} q_i(t) + \sigma(t) d_i \theta_i (1 - e^{-\Delta t/\theta_i})$$
$$+ \frac{[\sigma(t) - \sigma(t - \overline{\Delta t})] d_i \theta_i}{\overline{\Delta t}} [\Delta t - \theta_i (1 - e^{-\Delta t/\theta_i})] \qquad (3.17)$$

The determination of the value of the state variable $q(t + \Delta t)$ in an incremental process requires the knowledge of $q(t), \overline{\Delta t}, \sigma(t)$ and $\sigma(t - \overline{\Delta t})$. A similar algorithm is used in Abaqus. Other algorithms may be found in [7].

3.3 Physical Grounds and Extensions

Many inelastic properties of solids can be explained qualitatively in terms of various micro-structural rearrangements. In polymers, for example, the existence of long-chain molecules, which may straighten or crumple, in addition to sliding relatively to each other in response to sustained loads, provides the material with instantaneous elasticity as well as with nonlinear viscosity. During these processes, a certain amount of mechanical energy is lost into thermal energy. Additionally, the micro-structural changes give rise to macroscopic history-dependent material properties [2].

Explicit representations of history dependence may be formulated on purely mathematical assumptions and expressed as integrals of stress or strain history, as shown in Chap. 2, 7 and 8. Alternatively, a state variables representation may be used with some advantages in viscoelasticity as well as in plasticity and damage mechanics.

Then, the effect of the microscopic structural rearrangements is accounted for by the introduction of additional n state variables called internal variables or hidden coordinates, denoted collectively by $q_i(i = 1, \ldots, n)$ which, in a certain average global sense, represent the internal changes. The optimal selection of suitable internal variables, minimum in number, which provide maximum information, is an important problem [4].

Once the internal variables are chosen, we may define stress and internal energy as

$$\boldsymbol{\sigma} = \boldsymbol{\sigma}(\varepsilon, \Theta, q_i)$$
$$\Psi = \Psi(\varepsilon, \Theta, q_i)$$
(3.18)

where stress and free energy are expressed as functions of current values of strain (stress), temperature and other variables, including the internal state variables.

Rate effects are introduced through *evolution* or *growth laws*, in terms of the history of external quantities like the stress or strain tensors and temperature, as follows:

$$\dot{q}_i(t) = f(q_i, \Theta, \varepsilon)$$
(3.19)

From this set of equations one may explicitly eliminate the internal variables from the constitutive equations, thus obtaining a result similar to that from the functional theory in which stress (strain) is expressed as a functional of strain (stress).

An advantage of the state variable approach is that physical theories, and micro-structural information, may be introduced directly in the formulation of the evolution equations. Another one is that it leads to more efficient numerical procedures. This formulation will be extended to nonlinear viscoelasticity in Chap. 8. In the finite strain situations it is important to take account of rigid rotations and the related concepts of *State and Orientation* [5].

References

1. M.A. Biot, *Mechanics of Incremental Deformations* (Wiley, New York, 1965)
2. S. Nemat-Nasser, On Nonequilibrium Thermodynamics of Continua. in *Mechanics Today 2* (Pergamon Press, New York 1972)
3. E.T. Onat, Description of mechanical behavior of inelastic solids. in *Proceedings 5th US National Congress of Applied Mechanics*, 1966, pp. 421–434
4. E.T. Onat, The notion of state and its implications in thermodynamids of inelastic solids. in *Proceedings of IUTAM Symposium on Irreversible Aspects of Continuum Mechanics*, ed. by H. Pakus and L.I. Sedov (Springer, Berlin, 1968), pp. 292–314
5. E.T. Onat, Representation of inelastic mechanical behavior by means of state variables. in *proceedings of IUTAM Symposium on Thermoinelasticity*, ed. by B.A. Boley (Springer, 1970), pp. 213–224
6. J.C. Simo, T.J.R. Hughes, *Computational Inelasticity* (Springer, New York, 2000)
7. J. Sorvari, J. Hämäläinen, Time integration in linear viscoelasticity—a comparative study. Mech. Time-dependent. Mater **14**(3), 307–328 (2010)

Chapter 4
Multidimensional Viscoelastic Relations

In this chapter we extend the viscoelastic constitutive relations to a three dimensional setting. An important subject is the determination of parameters to be used in the solution of real problems. Viscoelastic behavior is determined experimentally, mainly through uniaxial creep tests. To obtain 3D relations, some simplifying assumptions are made. In the case of concrete, it is assumed that the Poisson ratio does not change with time. In the case of polymers, as the creep in shear is more important than volumetric creep, this one is disregarded and the material is considered as elastic in bulk. The viscoelastic equations are here presented in a general form and then simplified for isotropic materials. Finally, a procedure based on the state variables approach is presented for a general anisotropic linear viscoelastic material.

4.1 General Relations

In Chap. 2, we studied the viscoelastic constitutive relations in the one-dimensional case. In order to obtain the equations for the general case, we must proceed in the same way as in the elasticity theory, that is, we must substitute a relation between uniaxial stress σ and uniaxial strain ε by the corresponding relation between the tensors σ_{ij} and ε_{ij}. Thus, we define as a linear viscoelastic material one for which the relation between stresses and strains is given by

$$\sigma_{ij}(t) = \int_{\tau_0}^{t} E_{ijkl}(t - \tau)\dot{\varepsilon}_{kl}(\tau)d\tau \qquad (4.1)$$

where $E_{ijkl}(t - \tau)$ is a fourth order tensor relaxation function, or alternatively,

S. P. C. Marques and G. J. Creus, *Computational Viscoelasticity*,
SpringerBriefs in Computational Mechanics, DOI: 10.1007/978-3-642-25311-9_4,

$$\varepsilon_{ij}(t) = \int_{\tau_0}^{t} D_{ijkl}(t-\tau)\dot{\sigma}_{kl}(\tau)d\tau \tag{4.2}$$

where $D_{ijkl}(t-\tau)$ is a fourth order tensor creep function.

4.2 Isotropic Materials

4.2.1 Integral and Differential Representations

If the material is isotropic (i.e. the mechanical properties of an element are independent of its orientation) a simpler representation is possible because the tensor functions $E_{ijkl}(t,\tau)$ and $D_{ijkl}(t,\tau)$ have the same symmetries as the elastic tensor. Thus decomposing stress and strain history tensors in hydrostatic (σ_0, ε_0) and deviatoric (s_{ij}, e_{ij}) parts, as shown in Appendix B, we can write

$$s_{ij}(t) = \int_{\tau_0}^{t} 2G(t-\tau)\dot{e}_{ij}(\tau)d\tau$$
$$\sigma_0(t) = \int_{\tau_0}^{t} 3K(t-\tau)\dot{\varepsilon}_0(\tau)d\tau \tag{4.3}$$

or alternatively $\varepsilon_{ij}(t) = \varepsilon_0(t)\delta_{ij} + e_{ij}(t)$

$$e_{ij}(t) = \int_{\tau_0}^{t} D_G(t-\tau)\dot{s}_{ij}(\tau)d\tau$$
$$\varepsilon_0(t) = \int_{\tau_0}^{t} D_K(t-\tau)\dot{\sigma}_0(\tau)d\tau \tag{4.4}$$

Using the differential representation we have, extending (2.36),

$$\sum_{h=0}^{m} p_h^G \frac{\partial^h s_{ij}}{\partial t^h} = \sum_{k=0}^{n} q_k^G \frac{\partial^k e_{ij}}{\partial t^k}$$
$$\sum_{h=0}^{l} p_h^K \frac{\partial^h \sigma_0}{\partial t^h} = \sum_{k=0}^{r} q_k^K \frac{\partial^k \varepsilon_0}{\partial t^k} \tag{4.5}$$

which can be written in operator symbolic form as $P^G s_{ij} = Q^G e_{ij}$ and $P^K \sigma_0 = Q^K \varepsilon_0$, respectively, where

$$P^G = \sum_{h=0}^{m} p_h^G \frac{\partial^h}{\partial t^h} \qquad Q^G = \sum_{k=0}^{n} q_k^G \frac{\partial^k}{\partial t^k}$$

$$P^K = \sum_{h=0}^{l} p_h^K \frac{\partial^h}{\partial t^h} \qquad Q^K = \sum_{k=0}^{r} q_k^K \frac{\partial^k}{\partial t^k} \tag{4.6}$$

This differential representation will be explored using Laplace transform in Chap. 5.

The two material constants that appear in the constitutive relations of an isotropic elastic solid may be chosen as pairs (E, *v*), (K, G), etc. Relations among them may be found in Appendix B. Similar relations are valid, according to the *correspondence principle* (see Chap. 5), among the corresponding creep and relaxation functions and differential operators.

4.2.2 State Variables Representation

Let us consider equation (4.3)$_1$. We apply the procedure in Sect. 3.1 [see Eq. (3.7)] to the shear components

$$G(t - \tau) = G_\infty + \sum_{\alpha=1}^{n} G_\alpha e^{-(t-\tau)/T_\alpha} \tag{4.7}$$

Introducing the *n* quantities

$$q_{ij,\alpha}(t) = \int_{\tau_0}^{t} e^{-\frac{(t-\tau)}{T_\alpha}} \dot{e}_{ij}(\tau) d\tau \quad \alpha = 1, \ldots, n \tag{4.8}$$

and taking the derivative of (4.8) with respect to *t* (Leibnitz rule) we obtain

$$\dot{q}_{ij,\alpha}(t) + \frac{q_{ij,\alpha}(t)}{T_\alpha} = \dot{e}_{ij}(t) \quad \alpha = 1, \ldots, n \tag{4.9}$$

A combination of (4.3)$_1$, (4.7) and (4.8) provides the equation

$$s_{ij}(t) = 2G_\infty e_{ij}(t) + \sum_{\alpha=1}^{n} 2G_i q_{ij,\alpha}(t) \tag{4.10}$$

The same procedure is valid for the spherical components (4.3)$_2$ as well as for relations (4.4).

4.2.3 Determination of Creep and Relaxation Functions

Experimental determination of properties and curve fitting: A practical problem is the determination of the creep and relaxation functions in Eqs. (4.3) and

(4.4) from uniaxial rheological models or experimental results. The description of experimental procedures depends on the material addressed and is outside the scope of this book, but may be found in some other books [2–4] and in many papers. The experimental creep curve is usually approximated by an equation based on the Kelvin chain (see Sect. 2.3). Description of adequate curve fitting procedures may be found in [2]. For exponential series MATLAB curve fitting toolbox is useful. Commercial finite elements software, such as ABAQUS and ANSYS, do not use creep but relaxation function information. Procedures to transform creep into relaxation functions and vice versa are given in Sects. 2.4 and 5.1.

In many polymer materials, the creep function in shear is orders of magnitude higher than the corresponding bulk function. The experimental determination of the bulk viscoelasticity is difficult, and thus it is usual to model the material as elastic in bulk, with a K coefficient determined from the E and v values. In shear, the material is considered as viscoelastic, but usually the creep tests are performed not in shear but in axial tension or compression.

Procedures to obtain a shear relaxation function from an extensional relaxation creep function (assuming bulk elasticity) are given in Example 1 and, using *Laplace transforms*, in Sect. 5.1.

Example 1 In an example in the Abaqus Manual [1] (Viscoelastic rod subjected to constant axial load) it is used a Zener model (see Sect. 2.3, Example 1) for viscoelasticity. The extensional relaxation function of the bar material (a polymer) was obtained from tension tests as

$$E(t) = k_1 + k_2 e^{-t/T} \tag{4.11}$$

with $k_1 = 1000$, $k_2 = 9000$ and $T = 1$. To solve the problem with a finite element code (e.g. Abaqus) we need the elastic constants E and v, and the viscoelastic operators $G(t)$ and $K(t)$. It is assumed that the material is elastic in bulk, with a bulk coefficient $K = 100,000$. The initial elastic extensional modulus is $E(0) = k_1 + k_2$. The initial Poisson modulus is, from (4.11) and (B.4),

$$v(0) = \frac{3K - k_1 - k_2}{6K} = 0.4833 \tag{4.12}$$

The initial and final shear moduli are from Appendix B

$$G(0) = \frac{3K(k_1 + k_2)}{9K - (k_1 + k_2)} \quad G(\infty) = \frac{3Kk_1}{9K - k_1} \tag{4.13}$$

We need to determine $G(t)$. The relation among the viscoelastic operators is the same as the one among the corresponding elastic constants

$$G^* = \frac{3KE^*}{9K - E^*} \tag{4.14}$$

For the Zener material it is

$$E* = \left(\frac{1}{\eta} + \frac{\partial/\partial t(k_1 + k_2)}{k_1 k_2}\right)\left(\frac{1}{k_1\eta} + \frac{\partial/\partial t}{k_1 k_2}\right)^{-1} \tag{4.15}$$

with $\eta = k_2 T_E$. Substituting (4.15) into (4.14)

$$G(t) = \frac{3KE*}{9K - E*}H(t) = \left(3K\frac{\frac{1}{\eta} + \frac{\partial/\partial t(k_1+k_2)}{k_1 k_2}}{\frac{1}{k_1\eta} + \frac{\partial/\partial t}{k_1 k_2}}\right)\left(\frac{9K - \frac{1}{\eta} + \frac{\partial/\partial t(k_1+k_2)}{k_1 k_2}}{\frac{1}{k_1\eta} + \frac{\partial/\partial t}{k_1 k_2}}\right)^{-1}H(t)$$

$$\tag{4.16}$$

and simplifying

$$G(t) = \frac{3Kk_1 k_2}{9Kk_2 + 9K\eta\,\partial/\partial t - k_1 k_2 - \eta k_1\,\partial/\partial t - \eta k_2\,\partial/\partial t} \tag{4.17}$$

which leads to the differential equation

$$\dot{G}(t) + \frac{9Kk_2 - k_1 k_2}{9K\eta - \eta k_1 - \eta k_2}G(t) = \frac{3Kk_1 k_2}{9K\eta - \eta k_1 - \eta k_2} \tag{4.18}$$

whose solution, for the initial condition,

$$G(0) = \frac{3KE(0)}{9K - E(0)} = \frac{3Kk_1 + k_2}{9K - k_1 - k_2} \tag{4.19}$$

is, after substituting $\eta = k_2 T_E$

$$G(t) = \frac{27K^2 k_2}{(9K - k_1 - k_2)(9K - k_1)}e^{-\frac{(9K-k_1)t}{(9K-k_1-k_2)T_E}} + \frac{3Kk_1}{9K - k_1} \tag{4.20}$$

Then the relaxation time in shear is

$$T_G = -\frac{9K - k_1 - k_2}{9K - k_1}T_E = 0.9899 \tag{4.21}$$

For comparison, the same problem will be solved in Chapter 5 using Laplace transforms. These results will be used in Example 12.1.

4.3 Anisotropic Materials

4.3.1 Constitutive Relation for an Anisotropic Material

For the case of anisotropic materials, such as composites, the viscoelastic behavior is more complex and characterized by creep or relaxation constitutive tensors which present a greater number of independent component functions. In general, we can write the integral representation of an anisotropic linear viscoelastic material in the form given in (4.1) or (4.2). Using Voigt notation, these general constitutive relations can be written respectively as

$$\sigma_i(t) = \int_{\tau_0}^t E_{ij}(t - \tau) \frac{\partial \varepsilon_j(\tau)}{\partial \tau} d\tau$$

$$\varepsilon_i(t) = \int_{\tau_0}^t D_{ij}(t - \tau) \frac{\partial \sigma_j(\tau)}{\partial \tau} d\tau$$

$$\text{with} \quad i, j = 1, 2, \ldots, 6 \qquad (4.22)$$

where $E_{ij}(t)$ and $D_{ij}(t)$ represent the relaxation and creep functions of the material, respectively, $\varepsilon_i(t)$ are the components of strain and $\sigma_i(t)$ are the components of stress. Considering that the viscoelastic functions exhibit the same symmetry presented by the components of the elastic stiffness or elastic compliance i.e. $E_{ij}(t) = E_{ji}(t)$ and $D_{ij}(t) = D_{ji}(t)$, we conclude that in general an anisotropic viscoelastic material has 21 independent relaxation functions $E_{ij}(t)$ or creep functions $D_{ij}(t)$. This number is reduced in accordance to the material symmetry presented by the material. For example, that number is 9 for an orthotropic material and is 2 for an isotropic material.

In the next section, we present a general procedure based on the state variables approach for a general anisotropic linear viscoelastic material.

4.3.2 State Variables Representation

Upon integration by parts, Eq. (4.22)$_2$ may be written

$$\varepsilon_i(t) = D_{ij}(0) \sigma_j(t) - \int_0^t \frac{\partial}{\partial \tau} D_{ij}(t - \tau) \sigma_j(\tau) d\tau \qquad (4.23)$$

Approximating the creep functions by a Dirichlet–Prony series it can be written

$$D_{ij}(t - \tau) = D_{ij}^0 + \sum_{p=1}^M D_{ij}^p \left[1 - \exp\left(-\frac{t - \tau}{\theta_{ij}^p} \right) \right] \qquad (4.24)$$

where D_{ij}^0, D_{ij}^p and θ_{ij}^p are parameters to be determined from experimental results. M is the number of significant terms in the series and depends on the accuracy desired. The parameters θ_{ij}^p are the retardation times. Substituting (4.24) into (4.23)

$$\varepsilon_i(t) = D_{ij}(0) \sigma_j(t) + \sum_{p=1}^M \int_0^t d_{ij}^p(t - \tau) \sigma_j(\tau) d\tau \qquad (4.25)$$

where

$$d_{ij}^p = \frac{D_{ij}^p}{\theta_{ij}^p} \exp\left(-\frac{t-\tau}{\theta_{ij}^p}\right) \tag{4.26}$$

and no summation takes place on $i,\ j$. Thus,

$$\varepsilon_i(t) = D_{ij}(0)\sigma_j(t) + \sum_{p=1}^{M}\sum_{s=1}^{6} q_{is}^p(t) \tag{4.27}$$

where

$$q_{11}^p(t) = \int_0^t d_{11}^p(t-\tau)\sigma_1(\tau)\,d\tau$$

$$q_{12}^p(t) = \int_0^t d_{12}^p(t-\tau)\sigma_2(\tau)\,d\tau \tag{4.28}$$

$$\cdots$$

$$q_{66}^p(t) = \int_0^t d_{66}^p(t-\tau)\sigma_6(\tau)\,d\tau$$

are the state variables [5]. The numerical integration is as it was discussed in Chap. 3.

References

1. ABAQUS Theory manual. Information on the SIMULIA site: http://www.simulia.com/support/documentation.html
2. H.F. Brinson, L.C. Brinson, *Polymer Engineering Science and Viscoelasticity: An introduction* (Springer, New York, 2008)
3. R.M. Christensen, *Theory of Viscoelasticity*, 2nd edn. (Dover Publications, New York, 2010)
4. R.S. Lakes, *Viscoelastic Solids* (CRC Press LLC, Boca Raton, 1999)
5. S.P.C. Marques, G.J. Creus, Geometrically nonlinear finite elements analysis of viscoelastic composite materials under mechanical and hygrothermal loads. Comput. Struct. **53**, 449–456 (1994)

Chapter 5
Laplace Transform Solutions

Laplace Transform is a useful tool in solving important problems in different areas of science and engineering. Usually, it is employed to convert differential or integral equations into algebraic equations, simplifying the problem solutions. Particularly, in linear nonageing viscoelasticity, interesting applications have been found for Laplace transform techniques. Many computational solutions are also based on the use of Laplace transforms [10, 12]. As already mentioned, an important task in viscoelasticity consists of determining relations between the different constitutive viscoelasticity functions of a material [5, 9]. In this chapter, we show procedures based on Laplace transforms that allow us to obtain relaxation function given the corresponding creep function, or vice versa. Also, we show equivalence conditions between the integral and differential representations of the constitutive viscoelastic relations. In many practical situations, we know the creep function, which is evaluated in uniaxial tension or compression tests, and we need to determine the viscoelastic constitutive functions for multiaxial states of stress or strain. This problem is also focused in the present chapter. Finally, using the similarity between the mathematical formulations of the linear elastic and linear viscoelastic mechanical problems in the Laplace domain, the Correspondence Principle [2, 3] is stated and applied.

5.1 Relations Among Viscoelastic Constitutive Representations and Functions

Let $f(t)$ be a function of a real variable $t \geq 0$. The Laplace transform of $f(t)$ is defined by

$$L\{f(t)\} = \bar{f}(s) = \int_0^\infty e^{-st} f(t) dt \qquad (5.1)$$

S. P. C. Marques and G. J. Creus, *Computational Viscoelasticity*,
SpringerBriefs in Computational Mechanics, DOI: 10.1007/978-3-642-25311-9_5,
© The Author(s) 2012

where s is the transform parameter which may be complex or real. The reader can find more details about Laplace transform in, for instance, Myskis [4] and Wylie and Barrett [11]. The main Laplace transform properties used here are presented in Appendix A.

As we have seen earlier, the viscoelastic constitutive relations for an isotropic material can be defined in integral form by (4.3). Each one of these relations corresponds to a convolution of two functions (see Appendix A) as follows

$$s_{ij}(t) = 2G(t) \circ \frac{\partial e_{ij}(t)}{\partial t}$$

$$\sigma_0(t) = 3K(t) \circ \frac{\partial \varepsilon_0(t)}{\partial t} \tag{5.2}$$

Applying the Convolution theorem to (5.2), we find the following algebraic equations in the Laplace domain:

$$\bar{s}_{ij}(s) = 2s\bar{G}(s)\bar{e}_{ij}(s)$$

$$\bar{\sigma}_0(s) = 3s\bar{K}(s)\bar{\varepsilon}_0(s) \tag{5.3}$$

Similarly, for the constitutive relations (4.4), we have

$$\bar{e}_{ij}(s) = s\bar{D}_G(s)\bar{s}_{ij}(s)$$

$$\bar{\varepsilon}_0(s) = s\bar{D}_K(s)\bar{\sigma}_0(s) \tag{5.4}$$

Through simple algebraic manipulation of (5.3) and (5.4), the following equations involving Laplace transforms of the creep and relaxation functions are found:

$$2\bar{G}(s) = \frac{1}{s^2\bar{D}_G(s)} \tag{5.5}$$

$$3\bar{K}(s) = \frac{1}{s^2\bar{D}_K(s)} \tag{5.6}$$

These two equations are used to determine the relaxation functions $G(t)$ and $K(t)$ being given the corresponding creep functions $D^G(t)$ and $D^K(t)$, and vice versa. For the uniaxial extensional case, we find a similar relation connecting the Laplace transforms $\bar{D}(s)$ and $\bar{E}(s)$ of the creep function and relaxation function, respectively, which is given by

$$\bar{E}(s) = \frac{1}{s^2\bar{D}(s)} \tag{5.7}$$

The constitutive relations of a linear viscoelastic material can also be defined by the differential form given in (4.5). We now will show the connection among the coefficients of this differential representation and the creep and relaxation functions. The Laplace transforms of the terms of (4.5)$_1$ can be written as

$$L\{P^G s_{ij}\} = L\left\{s_{ij} + p_1^G \frac{\partial s_{ij}}{\partial t} + p_2^G \frac{\partial^2 s_{ij}}{\partial t^2} + \cdots + p_m^G \frac{\partial^m s_{ij}}{\partial t^m}\right\}$$

$$L\{Q^G e_{ij}\} = L\left\{q_0^G e_{ij} + q_1^G \frac{\partial e_{ij}}{\partial t} + q_2^G \frac{\partial^2 e_{ij}}{\partial t^2} + \cdots + q_n^G \frac{\partial^n e_{ij}}{\partial t^n}\right\}$$

(5.8)

Applying the rule of Laplace transform of derivatives (see Appendix A) and considering that the functions $s_{ij}(t)$ and $e_{ij}(t)$, as well as their derivatives with respect to time, vanish at $t \leq \tau_0$, the following equation in the Laplace domain can be found:

$$\bar{P}^G(s)\bar{s}_{ij}(s) = \bar{Q}^G(s)\bar{e}_{ij}(s) \tag{5.9}$$

where $\bar{P}^G(s)$ and $\bar{Q}^G(s)$ are polynomials in s associated to the differential operators P^G and Q^G, respectively, given by

$$\bar{P}^G(s) = p_0^G + p_1^G s + p_2^G s^2 + \cdots + p_m^G s^m$$

$$\bar{Q}^G(s) = q_0^G + q_1^G s + q_2^G s^2 + \cdots + q_n^G s^n$$

(5.10)

Similarly, for the constitutive relation defined by $(4.5)_2$, it can be shown that

$$\bar{P}^K(s)\bar{\sigma}_0(s) = \bar{Q}^K(s)\bar{\varepsilon}_0(s) \tag{5.11}$$

with

$$\bar{P}^K(s) = p_0^K + p_1^K s + p_2^K s^2 + \cdots + p_l^K s^l$$

$$\bar{Q}^K(s) = q_0^K + q_1^K s + q_2^K s^2 + \cdots + q_r^K s^r$$

(5.12)

Comparing (5.3) with (5.9) and (5.11) we obtain the relations

$$2\bar{G}(s) = \frac{1}{s}\frac{\bar{Q}^G(s)}{\bar{P}^G(s)} \tag{5.13}$$

$$3\bar{K}(s) = \frac{1}{s}\frac{\bar{Q}^K(s)}{\bar{P}^K(s)} \tag{5.14}$$

which state the equivalence conditions between the integral and differential representations of the constitutive viscoelastic relations.

Similarly, using (5.4), (5.9) and (5.11), we have

$$\bar{D}^G(s) = \frac{1}{s}\frac{\bar{P}^G(s)}{\bar{Q}^G(s)} \tag{5.15}$$

$$\bar{D}^K(s) = \frac{1}{s}\frac{\bar{P}^K(s)}{\bar{Q}^K(s)} \tag{5.16}$$

These last equations are alternative conditions to state the equivalence between the differential and integral forms.

Example 1 Determine the relaxation function in shear for a standard solid model being given its corresponding creep function as follows

$$D_G(t) = \frac{1}{2G_1} + \frac{1}{2G_2}\left[1 - \exp\left(-\frac{t}{\theta}\right)\right] \quad \text{with } \theta = \eta/G_2 \tag{5.17}$$

From Table A.1, the Laplace transform of (5.17) is given by

$$\bar{D}_G(s) = \left(\frac{1}{2G_1} + \frac{1}{2G_2}\right)\frac{1}{s} - \frac{1}{2G_2}\left(\frac{1}{\frac{1}{\theta}+s}\right)$$

and, substituting this equation in (5.5), we find the Laplace transform of the relaxation function

$$\bar{G}(s) = \frac{1 + \theta s}{s\left(\frac{1}{2G_1} + \frac{1}{2G_2} + \frac{\theta s}{2G_1}\right)} \tag{5.18}$$

Applying the partial fraction expansion technique [6], we write (5.18) as

$$\bar{G}(s) = \frac{A}{s} + \frac{B}{\left(\frac{1}{2G_1} + \frac{1}{2G_2} + \frac{\theta s}{2G_1}\right)} = \frac{A\left(\frac{1}{2G_1} + \frac{1}{2G_2}\right) + \left(A\frac{\theta}{2G_1} + B\right)s}{s\left(\frac{1}{2G_1} + \frac{1}{2G_2} + \frac{\theta s}{2G_1}\right)} \tag{5.19}$$

Comparing the second members of (5.18) and (5.19), the following expressions are obtained:

$$A\left(\frac{1}{2G_1} + \frac{1}{2G_2}\right) = 1 \quad A\frac{\theta}{2G_1} + B = \theta$$

and by solving this system of equations, we have

$$A = \frac{2G_1G_2}{G_1 + G_2}; \quad B = \frac{G_1\theta}{G_1 + G_2}$$

Substituting these equations in (5.19), we find the Laplace transform of the relaxation function

$$\bar{G}(s) = \frac{2G_1G_2}{G_1 + G_2} + \frac{2G_1^2}{G_1 + G_2}\left(\frac{1}{\frac{G_1+G_2}{G_2\theta}+s}\right) \tag{5.20}$$

Using Table A.1 to transform (5.20) to time domain, we find the following relaxation function in shear for the material:

$$G(t) = \frac{2G_1G_2}{G_1 + G_2} + \frac{2G_1^2}{G_1 + G_2}\exp\left(-\frac{t}{T}\right) \tag{5.21}$$

where $T = \eta/(G_1 + G_2)$. It is worth noting that the relaxation function corresponding to the pair (s_{ij}, e_{ij}) is $2G(t)$.

Example 2 Find the coefficients of the differential constitutive equation for a material that is elastic in dilatation and shows a viscoelastic behavior in shear, defined by a standard model, being given its relaxation function.

The differential constitutive relations for such a material can be defined by (see Sect. 4.2)

$$\sigma_0 = 3K\varepsilon_0 \quad \text{(in dilatation)}$$

$$p_0^G s_{ij} + p_1^G \frac{\partial s_{ij}}{\partial t} = q_0^G e_{ij} + q_1^G \frac{\partial e_{ij}}{\partial t}. \quad \text{(in shear)} \tag{5.22}$$

From (5.13), and applying the partial fraction expansion technique, we have

$$2\bar{G}(s) = \frac{q_0^G + q_1^G s}{s(1 + p_1^G s)} = \frac{A}{s} + \frac{B}{1 + p_1^G s} = \frac{A + (Ap_1^G + B)s}{s(1 + p_1^G s)}$$

which allows to conclude that

$$A = q_0^G \quad B = q_1^G - q_0^G p_1^G$$

Thus,

$$2\bar{G}(s) = \frac{q_0^G}{s} + \frac{q_1^G - q_0^G p_1^G}{1 + p_1^G s} = \frac{q_0^G}{s} + \frac{q_1^G - q_0^G p_1^G}{p_1^G} \frac{1}{\frac{1}{p_1^G} + s}$$

Using Table A.1 to transform this equation to time domain, we obtain

$$2G(t) = q_0^G + \left(\frac{q_1^G}{p_1^G} - q_0^G\right) \exp\left(-\frac{t}{p_1^G}\right) \tag{5.23}$$

Comparing (5.23) with (5.21), we have

$$p_1^G = T = \frac{\eta}{G_1 + G_2} \quad q_0^G = \frac{4G_1 G_2}{G_1 + G_2} \quad q_1^G = \frac{4G_1 \eta}{G_1 + G_2} \tag{5.24}$$

which, together with $p_0^G = 1$, are the coefficients of the differential constitutive representation in shear for the standard model.

Example 3 Consider an isotropic material whose behavior is elastic in bulk and viscoelastic in shear. The extensional axial creep function for this material is given by

$$D(t) = d_1 + d_2 \exp(-t/\theta) \tag{5.25}$$

Suppose that, in order to analyze a problem using a computer program (for instance, Abaqus), we need to provide the elastic constants E and v, as well as the viscoelastic functions $G(t)$ and $K(t)$. In this case, as the material is elastic in dilatation, $K(t)$ is constant and can be obtained directly as function of E and v (see

Appendix B). We need then to determine $G(t)$ from the known extensional axial creep function.

To solve this question we will initially obtain the extensional axial relaxation using the relation between creep function and relaxation function in the Laplace domain (Eq. (5.7)).

The Laplace transform of (5.25) is given by

$$\bar{D}(s) = \frac{d_1}{s} + d_2 \left(\frac{1}{\frac{1}{\theta} + s} \right)$$

Introducing this equation into (5.7), we have

$$\bar{E}(s) = \frac{1 + \theta s}{s[d_1 + (d_1 + d_2)\theta s)]}$$

whose inverse is the extensional uniaxial relaxation

$$E(t) = k_1 + k_2 \exp(-t/T) \tag{5.26}$$

where $k_1 = \dfrac{1}{d_1}$, $k_2 = -\dfrac{d_2}{d_1(d_1 + d_2)}$ and $T = \dfrac{d_1 + d_2}{d_1} \theta$.

Now, considering the material subjected to a uniaxial loading in direction 1 and using (2.5), we can write

$$\sigma_{11}(t) = \int_0^t E(t - \tau) \frac{\partial \varepsilon_{11}}{\partial \tau} d\tau \tag{5.27}$$

If the problem is decomposed in its spherical and deviator parts, we can easily obtain the relations

$$\sigma_{11}(t) = \int_0^t 3G(t - \tau) \frac{\partial e_{11}}{\partial \tau} d\tau \tag{5.28}$$

$$\sigma_{11}(t) = \int_0^t 9K(t - \tau) \frac{\partial \varepsilon_o}{\partial \tau} d\tau \tag{5.29}$$

where $e_{11} = 2(\varepsilon_{11} - \varepsilon_{22})/3$ and $\varepsilon_o = (\varepsilon_{11} + 2\varepsilon_{22})/3$. Using the Convolution theorem, the Laplace transforms of (5.27)–(5.29) are given, respectively, by

$$\bar{\sigma}_{11}(s) = s\bar{E}(s)\bar{\varepsilon}_{11}(s) \tag{5.30}$$

$$\bar{\sigma}_{11}(s) = 2s\bar{G}(s)[\bar{\varepsilon}_{11}(s) - \bar{\varepsilon}_{22}(s)] \tag{5.31}$$

$$\bar{\sigma}_{11}(s) = 3s\bar{K}(s)[\bar{\varepsilon}_{11}(s) + 2\bar{\varepsilon}_{22}(s)] \tag{5.32}$$

After a simple algebraic handling of (5.30)–(5.32), we find the relation

$$\bar{G}(s) = \frac{3\bar{K}(s)\bar{E}(s)}{9\bar{K}(s) - \bar{E}(s)} \tag{5.33}$$

This equation could be obtained directly by using the Correspondence Principle presented in Sect. 5.2.

Through (5.26) and considering $K(t) = K$, we have

$$\bar{E}(s) = \frac{k_1}{s} + k_2 \frac{1}{\frac{1}{T} + s} \quad \bar{K}(s) = \frac{K}{s} \tag{5.34}$$

Substituting (5.34) in (5.33), we obtain the relation

$$\bar{G}(s) = 3Kk_1 \frac{\frac{1}{T} + s}{s\left[\frac{9K - k_1}{T} + (9K - k_1 - k_2)s\right]} + 3Kk_2 \frac{1}{\frac{9K - k_1}{T} + (9K - k_1 - k_2)s} \tag{5.35}$$

Applying the partial fraction expansion technique, (5.35) can be written in the form

$$\bar{G}(s) = 3Kk_1 \left[\frac{A}{s} + \frac{B}{\left(\frac{9K - k_1}{T} + (9K - k_1 - k_2)s\right)}\right] + 3Kk_2 \frac{1}{\frac{9K - k_1}{T} + (9K - k_1 - k_2)s} \tag{5.36}$$

where $A = \dfrac{1}{9K - k_1}$ and $B = \dfrac{k_2}{9K - k_1}$. Using Table A.1, we obtain from (5.36) the following relaxation function in shear

$$G(t) = \frac{27K^2 k_2}{(9K - k_1 - k_2)(9K - k_1)} \exp\left(-\frac{9K - k_1}{9K - k_1 - k_2}\frac{t}{T}\right) + \frac{3Kk_1}{9K - k_1} \tag{5.37}$$

5.2 Correspondence Principle

In general, the mechanical behavior of a body must satisfy the motion equations, kinematic relations and constitutive equations. In this set of three types of different equations, only the constitutive relations are dependent on the material of which the body is made. Then, the mathematical formulation of a mechanical problem for a viscoelastic body is similar to that of an elastic body with the same geometry and subjected to identical boundary conditions, the only difference being the constitutive relations. The mechanical problem of a linear viscoelastic body with volume V and boundary surface S subjected to volume forces b_i, is defined by the following equations:

(a) Equilibrium equations (motion equations)

$$\frac{\partial \sigma_{ij}}{\partial x_j} + b_i = 0 \tag{5.38}$$

(b) Strain-displacement relations (kinematic equations)

$$\varepsilon_{ij} = \frac{1}{2}\left(\frac{\partial u_i}{\partial x_j} + \frac{\partial u_j}{\partial x_i}\right) \tag{5.39}$$

(c) Constitutive relations

$$P^G s_{ij} = Q^G e_{ij}$$
$$P^K \sigma_0 = Q^K \varepsilon_0 \tag{5.40}$$

(d) Boundary conditions

$$u_i = \widehat{u}_i \quad \text{on } S_u$$
$$\sigma_{ij} n_j = \widehat{t}_i \quad \text{on } S - S_u \tag{5.41}$$

where n_j are the components of the outward-directed unit vector normal to S_u with initial conditions

$$u_i = u_i^0 \quad \text{for } t = t^0 \tag{5.42}$$

The Laplace transforms of the equations given by the conditions (a)–(d) yield the following relations:

(a.1)

$$\frac{\partial \bar{\sigma}_{ij}(s)}{\partial x_j} + \bar{b}_i(s) = 0 \tag{5.43}$$

(b.1)

$$\bar{\varepsilon}_{ij}(s) = \frac{1}{2}\left(\frac{\partial \bar{u}_i(s)}{\partial x_j} + \frac{\partial \bar{u}_j(s)}{\partial x_i}\right) \tag{5.44}$$

(c.1)

$$\bar{P}^G(s)\bar{s}_{ij}(s) = \bar{Q}^G(s)\bar{e}_{ij}(s)$$
$$\bar{P}^K(s)\bar{\sigma}_0(s) = \bar{Q}^K(s)\bar{\varepsilon}_0(s) \tag{5.45}$$

(d.1)

$$\bar{u}_i(s) = \bar{\widehat{u}}_i(s) \quad \text{on } S_u$$
$$\bar{\sigma}_{ij}(s)n_j = \bar{\widehat{t}}_i(s) \quad \text{on } S - S_u \tag{5.46}$$

Comparing the conditions (a.1)–(d.1) with the governing field equations of linear elasticity, it can be concluded that they describe a fictitious quasi-static linear elastic problem defined by volume forces $\bar{b}_i(s)$, prescribed displacement

$\bar{u}_i(s)$ on S_u and external tractions $\bar{t}_i(s)$ on $S - S_u$ and elastic constants $\bar{Q}^G(s)/\bar{P}^G(s)$ and $\bar{Q}^K(s)/\bar{P}^K(s)$.

If the field variables $\bar{u}_i(s)$ and $\bar{\sigma}_{ij}(s)$ of this fictitious linear elastic problem can be obtained, then their inverse Laplace transforms $u_i(t)$ and $\sigma_{ij}(t)$ represent the displacement and stress fields of the viscoelastic problem, respectively. In other words, if the solution of the elastic problem is known, then the Laplace transform of the solution corresponding to the viscoelastic problem can be obtained by replacing the elastic constants $2G$ and $3K$ by the operator polynomial fractions $\bar{Q}^G(s)/\bar{P}^G(s)$ and $\bar{Q}^K(s)/\bar{P}^K(s)$, respectively, and the loads by their Laplace transforms. This analogy between elastic and viscoelastic problems is known as Correspondence Principle.

It is worth to notice that this principle cannot be used if the interface between the regions S_u and $S - S_u$ depends on time.

Example 4 Determine the viscoelastic operator polynomial fractions corresponding to Young modulus E and Poisson ratio v.

For an isotropic linear elastic material, one can obtain the relations

$$E = \frac{9KG}{3K + G} \qquad v = \frac{3K - 2G}{6K + 2G} \tag{5.47}$$

Introducing the operator polynomial fractions corresponding to G and K into these equations, the following relations can be found

$$E \rightarrow \frac{\dfrac{3}{2}\dfrac{\bar{Q}^K}{\bar{P}^K} \cdot \dfrac{\bar{Q}^G}{\bar{P}^G}}{\dfrac{\bar{Q}^K}{\bar{P}^K} + \dfrac{1}{2}\dfrac{\bar{Q}^G}{\bar{P}^G}} = \frac{3\bar{Q}^K\bar{Q}^G}{2\bar{Q}^K\bar{P}^G + \bar{Q}^G\bar{P}^K} \tag{5.48}$$

$$v \rightarrow \frac{\dfrac{\bar{Q}^K}{\bar{P}^K} - \dfrac{\bar{Q}^G}{\bar{P}^G}}{2\dfrac{\bar{Q}^K}{\bar{P}^K} + \dfrac{\bar{Q}^G}{\bar{P}^G}} = \frac{\bar{P}^G\bar{Q}^K - \bar{P}^K\bar{Q}^G}{2\bar{P}^G\bar{Q}^K + \bar{P}^K\bar{Q}^G} \tag{5.49}$$

For the sake of simplification the variable s was omitted in these relations.

Example 5 Using Laplace transformation method, to determine the solution of the problem consisting of a linear viscoelastic thick-walled long cylinder subjected to an internal pressure p as shown in Fig. 5.1. The material is assumed as elastic in dilatation and viscoelastic in shear with a behavior defined by (a) standard solid model and (b) Zener model.

Considering plane strain state and linear elastic material, the stress and radial displacement fields are found by the theory of elasticity as follows [7]

$$\sigma_r = \frac{pb^2}{a^2 - b^2}\left(1 - \frac{a^2}{r^2}\right) \tag{5.50}$$

$$\sigma_\theta = \frac{pb^2}{a^2 - b^2}\left(1 + \frac{a^2}{r^2}\right) \tag{5.51}$$

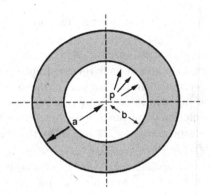

$$u = \frac{(1+v)pb^2}{E(a^2-b^2)}\left[(1-2v)r+\frac{a^2}{r}\right] \tag{5.52}$$

Substituting the expressions (5.47) into (5.52), we find the following equivalent
equation for the radial displacements

$$u = \frac{pb^2}{2G(a^2-b^2)}\left[\frac{3G}{3K+G}r+\frac{a^2}{r}\right] \tag{5.53}$$

As the radial and transversal stresses do not depend on the elastic constants, the
equations (5.50) and (5.51) are also valid for the viscoelastic cylinder. Replacing 2G
and 3 K by their corresponding polynomial fractions $\bar{Q}^G(s)/\bar{P}^G(s)$ and
$\bar{Q}^K(s)/\bar{P}^K(s)$, respectively, in (5.53), we obtain the Laplace transform for the radial
displacement of the associated viscoelastic problem as follows

$$\bar{u} = \frac{\bar{p}b^2}{(a^2-b^2)}\frac{\bar{P}^G}{\bar{Q}^G}\left(\frac{3\bar{P}^K\bar{Q}^G}{2\bar{P}^G\bar{Q}^K+\bar{P}^K\bar{Q}^G}r+\frac{a^2}{r}\right) \tag{5.54}$$

If the material is linear elastic in dilatation and viscoelastic in shear with behavior
described by a standard solid model or Zener model, its polynomials can be written as

$$\bar{P}^K(s) = 1 \quad \bar{Q}^K(s) = 3K$$
$$\bar{P}^G(s) = 1+p_1^Gs \quad \bar{Q}^G(s) = q_o^G+q_1^Gs \tag{5.55}$$

Assuming $p(t) = pH(t)$, its Laplace transform is $\bar{p}(s) = \frac{p}{s}$. Substituting the
polynomials (5.55) in (5.54), we obtain

$$\bar{u} = \frac{pb^2}{(a^2-b^2)}\left\{\frac{3(1+p_1^Gs)r}{s\left[6K+q_0^G+(6Kp_1^G+q_1^G)s\right]}+\frac{a^2}{r}\frac{1+p_1^Gs}{s(q_0^G+q_1^Gs)}\right\} \tag{5.56}$$

Through the partial fraction expansion technique, (5.56) can put in the form

$$\bar{u} = \frac{pb^2}{(a^2 - b^2)} \left\{ 3r \left[\frac{A}{s} + \frac{B}{6K + q_0^G + (6Kp_1^G + q_1^G)s} \right] + \frac{a^2}{r} \left(\frac{C}{s} + \frac{D}{q_0^G + q_1^G s} \right) \right\} \tag{5.57}$$

where

$$A = 1/(6K + q_o^G), \quad B = (q_o p_1^G - q_1^G)/(6K + q_o^G),$$
$$C = 1/q_o^G, \quad D = p_1^G - q_1^G/q_o^G.$$

The solution for the radial displacements of the viscoelastic cylinder can be easily obtained using Table A.1 to invert the Laplace transform (5.57). Hence, the following general equation for the radial displacements for both models is found

$$u = \frac{pb^2}{(a^2 - b^2)} \left\{ 3r \left[\frac{1}{6K + q_o^G} + \left(\frac{p_1^G}{6Kp_1^G + q_1^G} - \frac{1}{6K + q_o^G} \right) \exp\left(-\frac{6K + q_o^G}{6Kp_1^G + q_1^G} t \right) \right] \right.$$
$$\left. + \frac{a^2}{q_o^G r} \left[1 + \left(\frac{q_o^G}{q_1^G} p_1^G - 1 \right) \exp\left(-\frac{q_o^G}{q_1^G} t \right) \right] \right\} \tag{5.58}$$

Introducing the parameters of the considered model into (5.58), we have the radial displacement solution. The parameters of the standard model are given in (5.24), with $p_0^G = 1$, and those for the Zener model are

$$p_o^G = 1 \quad p_1^G = \frac{\eta}{G_2} \quad q_o^G = 2G_1 \quad q_1^G = \frac{2\eta}{G_2}(G_1 + G_2) \tag{5.59}$$

From the limit theorems presented in Appendix A we can obtain the radial displacement $u(0)$ and $u(\infty)$ using the Laplace transform $\bar{u}(s)$ given in (5.56). For instance, the second limit theorem states that $\lim_{s \to 0} s\bar{u}(s) = u(\infty)$. Applying this result to (5.56), we can easily obtain

$$u(\infty) = \frac{pb^2}{(a^2 - b^2)} \left(\frac{3r}{6K + q_o^G} + \frac{a^2}{q_o^G r} \right) \tag{5.60}$$

that is in accordance with (5.58).

5.3 Numerical Inversion of Laplace Transform

In problems presented above we have used an analytical procedure, the partial fraction expansion method, to invert the Laplace transforms. However, in many cases of practical interest that inversion cannot be easily obtained and, then, we need to use numerical tools to do it. This can be considered as a weakness of the Laplace transform techniques. On the other hand, in many problems the elastic solutions are known only in a numerical form and, for these cases, the application of the correspondence principle requires the use of numerical tools to obtain their associated viscoelastic solutions.

In this section, we present a numerical procedure to obtain the inverse function of a Laplace transform, which is particularly appropriate to apply in linear viscoelasticity problems when the analytical solutions for that inversion are not available. This procedure was firstly presented by Schapery [8] and is known as the Collocation Method.

Suppose the function $f(t)$ in (5.1) can be approximate by a Dirichlet-Prony series with N terms, as follows,

$$f_A(t) = \sum_{j=1}^{N} X_j \exp\left(-\frac{t}{t_j}\right) \tag{5.61}$$

where X_j are unknown coefficients and t_j are discrete times conveniently selected. $f_A(t)$ indicates an approximation for the function $f(t)$.

The total square error between those functions is defined by

$$\Delta = \int_0^\infty [f(t) - f_A(t)]^2 dt \tag{5.62}$$

To minimize this total square error with respect to the unknown coefficients in (5.61), we impose the following N conditions

$$\frac{\partial \Delta}{\partial X_i} = \int_0^\infty 2[f(t) - f_A(t)]\left[-\frac{\partial f_A(t)}{\partial X_i}\right] dt = 0 \quad \text{with } i = 1, 2, \ldots N \tag{5.63}$$

Introducing (5.61) into (5.63), we obtain

$$\int_0^\infty f(t) \exp\left(-\frac{t}{t_i}\right) dt - \sum_{j=1}^{N} X_j \int_0^\infty \exp\left[-\left(\frac{1}{t_i} + \frac{1}{t_j}\right)t\right] dt = 0$$

or, noting that the first term is the Laplace transform of $f(t)$ at $s = 1/t_i$ e solving the second integral,

$$\sum_{j=1}^{N} X_j \frac{t_i t_j}{t_i + t_j} = \bar{f}(s)\Big|_{s=\frac{1}{t_i}} \tag{5.64}$$

Using indicial notation, (5.64) can be compactly written as

$$K_{ij} X_j = \bar{f}_i \tag{5.65}$$

where

$$K_{ij} = \frac{t_i t_j}{t_i + t_j}; \quad \bar{f}_i = \bar{f}(s)\Big|_{s=\frac{1}{t_i}} \tag{5.66}$$

Equation (5.65) represents a linear set of N equations with N unknowns (X_j). With the N values t_k selected and with the values of the Laplace transform of $f(t)$ for $s = 1/t_k$, the coefficients appearing in (5.61) can be determined by solving the linear system of equations (5.65) and, then, we have the approximated function $f_A(t)$. An alternative approach to the above procedure can be found in Barbero [1].

References

1. E.J. Barbero, *Finite Element Analysis of Composite Materials* (CRC Press, Boca Raton, 2008)
2. E.H. Lee, Stress analysis in viscoelastic bodies. Q. Appl. Math. **13**, 183–190 (1955)
3. L.W. Morland, E.H. Lee, Stress analysis for linear viscoelastic materials with temperature variation. Trans. Soc. Rheol. **4**, 223–263 (1960)
4. A.D. Myskis, *Advanced Mathematics for Engineers* (Mir Publishers, Moscow, 1979)
5. S.W. Park, R.A. Schapery, Methods of interconversion between linear viscoelastic material functions: part I—A numerical method based on Prony series. Int. J. Solids Struct. **36**, 1653–1675 (1999)
6. K.F. Riley, M.P. Hobson, S.J. Bence, *Mathematical Methods for Physics and Engineering* (Cambridge University Press, New York, 2006)
7. M.H. Sadd, *Elasticity: Theory, Applications and Numerics* (Elsevier Butterworth-Heinemann, Burlington, 2005)
8. R.A. Schapery, Approximate methods of transform inversion for viscoelastic stress analysis. Proceedings of 4th U.S. National Congress of Applied Mechanics, ASME, 1962, p. 1075
9. R.A. Schapery, S.W. Park, Methods for interconversion between linear viscoelastic material functions: part II—an approximate analytical method. Int. J. Solids Struct. **36**, 1677–1699 (1999)
10. M. Schanz, H. Antes, T. Rüberg, Convolution quadrature boundary element method for quasi-static visco- and poroelastic continua. Comput. Struct. **83**, 673–684 (2005)
11. C.R. Wylie, L.C. Barrett, *Advanced Engineering Mathematics* (McGraw-Hill, New York, 1995)
12. Y. Yeong-Moo, P. Sang-Hoon, Y. Sung-Kie, Asymptotic homogenization of viscoelastic composites with periodic microstructures. Int. J. Solids Struct. **35**, 2039–2055 (1998)

Chapter 6
Temperature Effect

The viscoelastic constitutive relations presented so far were developed under the hypothesis of isothermal conditions. However, most viscoelastic materials, particularly polymers, have temperature dependent constitutive relations. The mechanisms responsible for these thermal effects have micro-structural origin and are, consequently, complex. In this chapter we present a brief description on temperature effects on the linear viscoelasticity behavior of polymers and concrete and a simplified formulation that is adequate for the so called thermo-rheologically simple materials.

6.1 Linear Thermoviscoelasticity

In the following, we use Voigt notation for tensors. The total strain of a solid subjected simultaneously to a mechanical loading and a temperature change can be assumed as composed by two parts, as follows:

$$\varepsilon_i(t) = \varepsilon_i^M(t) + \varepsilon_i^T(t) \tag{6.1}$$

where ε_i^M and ε_i^T are the mechanical and thermal strains, respectively. In this case, for a linear viscoelastic solid, the constitutive relation can be written as

$$\varepsilon_i(t) = \int_{\tau_0}^{t} D_{ij}(t - \tau, \Theta) \frac{\partial \sigma_j(\tau)}{\partial \tau} d\tau + \varepsilon_i^T(t) \tag{6.2}$$

or, alternatively,

$$\sigma_i(t) = \int_{\tau_0}^{t} E_{ij}(t - \tau, \Theta) \frac{\partial \left[\varepsilon_j(\tau) - \varepsilon_j^T(\tau) \right] (\tau)}{\partial \tau} d\tau \tag{6.3}$$

S. P. C. Marques and G. J. Creus, *Computational Viscoelasticity*,
SpringerBriefs in Computational Mechanics, DOI: 10.1007/978-3-642-25311-9_6,
© The Author(s) 2012

being Θ the temperature, which can be constant or variable with time. Then, the creep and relaxation functions depend on both time and temperature. The thermal strain component is given by

$$\varepsilon_i^T(t) = \alpha_i \Delta\Theta(t) \qquad (6.4)$$

where, in general, the thermal expansion coefficients of the material α_i, $(i = 1, 2, ..., 6)$, can be represented as the components of the vector $\boldsymbol{\alpha} = \{\alpha_{11}, \alpha_{22}, \alpha_{33}, \alpha_{12}, \alpha_{13}, \alpha_{23}\}$. For the principal directions of an orthotropic material, the last three components of this vector are null, meaning that the temperature causes only normal strains. In case of isotropic materials, $\alpha_{11} = \alpha_{22} = \alpha_{33} = \alpha$ and $\alpha_{12} = \alpha_{13} = \alpha_{23} = 0$. In (6.4), $\Delta\Theta$ is the temperature variation with respect to the initial strain-free temperature.

The thermal expansion coefficients may also depend on temperature Θ. For most materials, subjected to temperature in the usual range of structural applications, coefficients α_i can be assumed as approximately linear functions of temperature Θ [10]. In this case the thermoviscoelastic formulation is nonlinear.

6.2 Temperature Effects in Polymers

The thermoviscoelastic behavior of polymers is related to molecular rearrangements under stress whose speed depends on temperature [11]. In general, the polymers may present different molecular transitions, the most important being the rubber-glass transition defined by the glass-transition temperature T_g.

At a temperature above T_g an amorphous polymer exhibits high rates of deformation, behaving like a rubber, i.e. presenting large, practically instantaneous and fully reversible strains when subjected to mechanical loads. On the other hand, at temperature below T_g the polymer presents low deformation. In this case, the material behaves like a glass, exhibiting instantaneous and reversible strain and brittle fracture. For a range of intermediate temperatures near T_g the polymer has a behavior that consists of a combination of those exhibited in the glassy and rubbery regimes.

Figure 6.1 shows a typical curve that represents the variation of a relaxation modulus E, for an amorphous polymer, versus temperature in the three cases. As shown in this figure, for the glassy and rubbery regimes the modulus is not sensitive to temperature variations while in the viscoelastic regime it is strongly dependent on temperature.

Each molecular transition is associated with a relaxation mechanism. Some materials exhibit only one dominant molecular transition: these are *thermorheologically simple materials*. Many amorphous polymers behave as *thermorheologically simple* [18]. For these materials, a temperature change results in an horizontal shift of the viscoelastic function (creep or relaxation) when plotted against log t as abscissa (see Fig. 6.2). On the other hand, some materials present

Fig. 6.1 Typical variation of a polymer relaxation modulus versus temperature

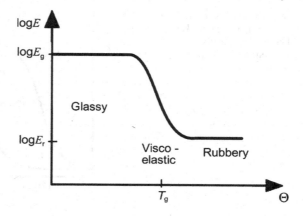

two or more molecular transitions and relaxation mechanisms. Their behavior is *thermorheologically complex.* More details about polymer properties and behavior can be found in Ferry [9], Callister Jr [8], Brinson and Brinson [7].

6.3 Thermorheologically Simple Materials

To verify if a material belongs to this class, we must perform a set of creep or relaxation tests at different temperatures and plot the results using a logarithm time scale. If the curves can be superposed by a horizontal shift to form a master curve, the material is considered thermorheologically simple [12]. This procedure is very useful to obtain long term results using short term tests.

6.3.1 Time-Shifting Factor

Let us consider the creep curves in Fig. 6.2. At a reference temperature Θ_0 we have a creep function $D(t - \tau, \Theta_0)$. Making a change of variables, i.e., using a decimal logarithmic scale for time, the creep function can be written as

$$D(t - \tau, \Theta_0) = L(\log(t - \tau)) \tag{6.5}$$

and, for temperature Θ,

$$D(t - \tau, \Theta) = L(\log(t - \tau) - \varphi(\Theta)) \tag{6.6}$$

with $\varphi(\Theta_0) = 0$ and $\partial\varphi(\Theta)/\partial\Theta < 0$. Writing $\varphi(\Theta) = \log a_T(\Theta)$, we have now

Fig. 6.2 Creep curves at two different temperatures for a thermorheologically simple material; a_T is the time shift factor

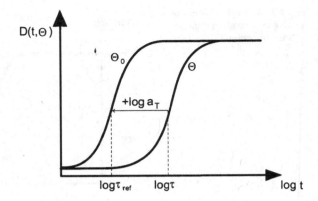

$$D(t - \tau, \Theta) = L\left(\log \frac{t - \tau}{a_T(\Theta)}\right) = L(\log(\xi - \xi')) = D(\xi - \xi', \Theta_0) \qquad (6.7)$$

where $\xi = t/a_T(\Theta)$ and $\xi' = \tau/a_T(\Theta)$ are the *reduced times*. The function $a_T(\Theta)$ is called *shift factor* [19].

The relation between real and reduced time, i.e., $a_T(\Theta)$, is obtained experimentally from creep or relaxation tests as shown in Fig. 6.2. For temperature near or above the glass transition temperature, Williams et al. [19] proposed an empirical formula, known as WLF (William–Landel–Ferry) equation

$$\log a_T = \frac{-C_1(\Theta - \Theta_{ref})}{C_2 + \Theta + \Theta_{ref}} \qquad (6.8)$$

where C_1 and C_2 are material constants dependent on the chosen reference temperature Θ_{ref}. When $\Theta_{ref} = T_g$, it has been proposed that C_1 and C_2 assume the "universal" values $C_1 = 17.4$ and $C_2 = 51.6$, which are applicable to a wide range of polymers.

For the case of transient temperature conditions, Morland and Lee [14] proposed the following equation for the reduced time:

$$\xi(t) = \int_0^t \frac{1}{a_T(\Theta(\tau))} d\tau \qquad (6.9)$$

where τ is an arbitrary real variable in the interval $[0, t]$.

6.3.2 Real Time Behavior

We saw in Chap. 3 that the behavior of a linear viscoelastic material may be represented by a Kelvin chain. Thus, consider a Kelvin element for which the creep function at a reference temperature Θ_0 is given by (2.21)

$$D(t - \tau, \Theta_0) = \frac{1}{E}\left[1 - \exp\left(-\frac{t - \tau}{\theta}\right)\right]; \quad \theta = \eta/E \tag{6.10}$$

For a temperature Θ, in accordance with (6.7), the creep function can be written as

$$D(t - \tau, \Theta) = L(\log(t - \tau) - \varphi(\Theta)) = \frac{1}{E}\left[1 - \exp\left(-\frac{\xi - \xi'}{\theta}\right)\right] \tag{6.11}$$

where $\xi = t/a_T(\Theta)$ and $\xi' = \tau/a_T(\Theta)$ are the reduced times. Using the real time, we have then

$$D(t - \tau, \Theta) = \frac{1}{E}\left[1 - \exp\left(-\frac{t - \tau}{\hat{\theta}}\right)\right] \tag{6.12}$$

with $\hat{\theta} = a_T(\Theta)\theta$. This situation corresponds to a Kelvin element with

$$E(\Theta) = E(\Theta_0) = \text{constant}$$

$$\eta(\Theta) = a_T(\Theta)\eta(\Theta_0)$$

This example shows the characteristics of the hypothesis of thermo-rheological simplicity:

(1) time shifting modifies retardation or relaxation times, but not the value of the final deformations;
(2) the procedure is still valid when we have more than one element in the Kelvin chain (or Maxwell chain) as long as all of them depend in the same form on temperature, but if temperature dependence is different in each element, the response is not thermo-rheologically simple. Thus, materials in which the relaxation time is dominated by one thermally activated process will be thermorheologically simple, but if there are multiple relaxation mechanisms, each with its own dependence on temperature, the material will not be thermorheologically simple [16, 17].
(3) the thermorheologically simple condition implies that the material dependence on temperature is governed only by the single temperature function $a_T(\Theta)$ [15].

6.3.3 Anisotropic Materials

For an anisotropic thermorheologically simple material, the viscoelastic functions and consequently the reduced times, depend on the direction. In this case, the creep functions at a base temperature Θ_0 may be in general written in the form

$$D_{ij}(t - \tau, \Theta_0) = L_{ij}(\log(t - \tau)) \tag{6.13}$$

and for a temperature Θ

$$D_{ij}(t - \tau, \Theta) = L_{ij}\big(\log(t - \tau) - \varphi_{ij}(\Theta)\big) \tag{6.14}$$

with $\varphi_{ij}(\Theta_0) = 0$ and $\partial\varphi_{ij}(\Theta)/\partial\Theta < 0$. Introducing shift factors $a_{ij}^T(\Theta)$, we have

$$D_{ij}(t - \tau, \Theta) = D_{ij}(\xi_{ij} - \xi'_{ij}, \Theta_0) \tag{6.15}$$

being $\xi_{ij} = t\big/a_{ij}^T(\Theta)$ and $\xi'_{ij} = \tau\big/a_{ij}^T(\Theta)$ reduced times.

6.4 Thermo-Rheologically Complex Materials

As mentioned in Sect. 6.2, the polymer materials can present different molecular transitions which correspond to distinct characteristic temperature dependence. When the polymer exhibits a single molecular process, it behaves as a thermo-rheologically simple material (Sect. 6.3). In contrast, if the polymer presents two or more active molecular processes, each one of them is associated with a different shifting implying a thermorheologically complex behavior. In this case, the superposition or fit of viscoelastic functions at different temperature values, plotted against time logarithm scale, usually requires horizontal and vertical shifts and eventually rotations. Barbero [3] presents a methodology to predict long-term creep of polymeric composite laminates from short-term constituent data from which the matrix is modeled as a thermorheologically complex material with time-dependent horizontal and vertical shift factors. This model is validated by experimental results.

A procedure also used to describe the thermorheologically complex behavior of linear viscoelastic polymers consists of separating the contributions corresponding to the different molecular transitions. In this context, the polymer behavior can be represented by assemblages of springs and dashpots connected in series or in parallel, like those generalized models shown in Chap. 2, considering separated contributions from the molecular processes. If we indicate by α and β two distinct molecular transitions, the generalized Maxwell model shown in Fig. 6.3 can be used to represent the polymer behavior [11]. This rheological model is constituted of $n + m$ Maxwell element units, where n and m indicate the number of these units referred to the molecular transition α and β, respectively. A generalized Kelvin representation can also be employed to model the same complex behavior (see [11].

For the generalized Maxwell model shown in Fig. 6.3, the relaxation function is defined by

$$E(t) = \sum_{i=1}^{m} E_{\alpha,i} \exp\left(-\frac{t}{T_{\alpha,i}}\right) + \sum_{j=1}^{n} E_{\beta,j} \exp\left(-\frac{t}{T_{\beta,j}}\right) \tag{6.16}$$

where the relaxation times $T_{\alpha,i} = \eta_{\alpha,i}/E_{\alpha,i}$ and $T_{\beta,j} = \eta_{\beta,j}/E_{\beta,j}$.

Fig. 6.3 Generalized
Maxwell model for a
thermorheologically complex
material

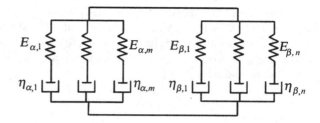

6.5 Temperature Effects in Concrete

Concrete is a material that at work stresses behaves as linear viscoelastic. However, its behavior is complex because of two additional effects: aging and
shrinkage. Creep in concrete can be divided into *basic creep* and *drying creep*.
Basic creep occurs without moisture movement to or from the environment
whereas *drying creep* is caused by drying. The primary mechanism responsible for
the creep and relaxation of concrete is considered as the activation energy in
cement paste. Long-term applied stress exceeds the activation energy limit of the
material and triggers the breaking of the bond in cement paste which leads to long-
term deformation [4]. At room temperature conditions and low stress level, the
creep strain for a stress σ_0 applied at time τ_0 can be evaluated by

$$\varepsilon(t, \tau_0) = D(t, \tau_0)\sigma_0 = \frac{1}{E_0}[1 + \phi(t, \tau_0)]\sigma_0 \qquad (6.17)$$

where $D(t, \tau_0)$ indicates the concrete creep function, $\phi(t, t_0)$ is the creep coefficient
and E_0 is the reference Young modulus. In this case, the creep function does not
depend on stress and thus, the strain at time t is proportional to the value of σ_0.
Under this condition, the linear viscoelasticity theory may be applied. In the
current design practice, such linearity is assumed for stress levels below 50% of
the strength limit.

Usually, temperature accelerates the creep process in concrete. When the
temperature level is not too high, it is possible that the material creep mechanism
continues being that one corresponding to room temperature. In this situation, the
creep function depends on temperature and the creep process is accelerated, but its
mechanism is not altered. So, according to Bazant and Kaplan [4], the same creep
function obtained at room temperature could be used together with an accelerated
time scale derived from Arrhenius equation. This means that under such conditions
of temperature concrete obeys a time–temperature equivalence relation. However,
for concrete, in general, a temperature variation implies horizontal and vertical
shifts, as well as rotation, of the creep curve [2]. Consequently, concrete does not
behave as a thermorheologically simple material.

In contrast, for higher temperatures the creep mechanism is changed and the
creep function becomes dependent on both temperature and stress. Under this

elevated temperature conditions, the stress dependency of the creep function may occur even in very low stress levels and the idea of using an accelerated time scale applied to the room temperature creep function cannot be employed. More details about creep in concrete may be found in Arthananari and Yu [1], Marechal [13] and Bažant and Kim [5, 6].

References

1. S. Arthananari, C.W. Yu, Creep of concrete under uniaxial and biaxial stresses at elevated temperatures. Mag. Concr. Res. **19**(60), 149–156 (1967)
2. O.R. Barani, D. Mostofinejaad, M.M. Saadatpour, M. Shekarchi, Concrete basic creep prediction based on time–temperature equivalence relation and short-term tests. Arab. J. Sci. Eng. **35**(2B), 105–121 (2010)
3. E.J. Barbero, Prediction of long-term creep of composites from doubly-shifted polymer creep data. J. Compos. Mater. **43**(19), 2109–2124 (2009)
4. Z.P. Bazant, M.F. Kaplan, *Concrete at High Temperatures: Material Properties and Mathematical Models* (Longman Group Limited, Longman House, Burnt Mill, Harlow, 1996)
5. Z.P. Bažant, J.K. Kim, Improved prediction model for time-dependent deformations of concrete: part 2—basic creep. Mater. Struct. **24**, 409–421 (1991)
6. Z.P. Bažant, J.K. Kim, Improved prediction model for time-dependent deformations of concrete: part 4—temperature effects. Mater. Struct. **25**, 84–94 (1992)
7. H.F. Brinson, L.C. Brinson, *Polymer Engineering Science and viscoelasticity: An Introduction* (Springer, New York, 2008)
8. W. Callister Jr, *Materials Science and Engineering: An Introduction* (Wiley, New York, 2003)
9. J.D. Ferry, *Viscoelastic Properties of Polymers*, 3rd edn. (Wiley, New York, 1980)
10. W.N. Findley, J.S. Lai, K. Onaran, *Creep and Relaxation of Nonlinear Viscoelastic Materials* (Dover Publications Inc., New York, 1989)
11. E.T.J. Klompen, L.E. Govaert, Nonlinear viscoelastic behaviour of thermorheologically complex materials. Mech. Time Depend. Mater. **3**, 49–69 (1999)
12. R.S. Lakes, *Viscoelastic Solids* (CRC Press LLC, Boca Raton, 1999)
13. J.C. Marechal, Creep of concrete as a function of temperature. In: *International Seminar on Concrete for Nuclear Reactors*, ACI Special Publication No. 34, vol. 1, American Concrete Institute, Detroit, 547–564 (1972)
14. L.W. Morland, E.H. Lee, Stress analysis for linear viscoelastic materials with temperature variation. Trans. Soc. Rheol. **4**, 223 (1960)
15. R. Muki, E. Sternberg, On transient thermal stresses in viscoelastic materials with temperature-dependent properties. J. Appl. Mech. **28**, 193–207 (1961)
16. D.J. Plazek, Temperature dependence of the viscoelastic behavior of polysterene. J. Phys. Chem. **69**, 3480–3487 (1965)
17. D.J. Plazek, Oh, thermorheologically simplicity, wherefore art thou? J. Rheol. **40**, 987–1014 (1996)
18. F. Schwarzl, A.J. Starveman, Time-temperature dependent of linear viscoelastic behavior. J. Appl. Phys. **23**, 838–843 (1952)
19. M.L. Williams, R.F. Landel, J.D. Ferry, The temperature dependence of relaxation mechanisms in amorphous polymers and other glass-forming liquids. Temp. Dependence Relax. Mech. **77**, 3701–3707 (1955) (Contribution from the Department of Chemistry, University of Wisconsin)

Chapter 7
Materials with Aging

We call aging the change in the mechanical properties of a given material with *age* which is the time period between some origin more or less arbitrarily established and the time of observation. Concrete is a material that may be used as an example: from the moment of casting (taken usually as age zero) it begins to increase its strength and to decrease its deformability. In the case of polymers both physical (reversible) and chemical (irreversible) aging are observed. In the present chapter we introduce the equations for viscoelasticity with aging for situations in which compliance diminishes ("hardening") and for situations in which compliance increases ("softening") with age in integral form and through rheological models and state variables equations. The time-age equivalence model applied to the physical aging of polymers is also discussed.

7.1 Experimental Results

The function $D(t, \tau)$ for concrete has the form indicated in Fig. 1.5. The aging of concrete may be classified as chemical because it is due to a progressive hydration process during a period of several months after casting and is irreversible [2, 5]. Figure 7.1 shows results for physical aging of a polymer, which is temperature driven and reversible. PVC specimens were quenched from 90 to 40°C and tested at the lower temperature at different times.

Sometimes, the concept of aging involves other influences in addition to time. Environmental conditions are also important: for concrete, humidity and temperature; for polymers, temperature, humidity and UV radiation in the case of chemical aging.

S. P. C. Marques and G. J. Creus, *Computational Viscoelasticity*,
SpringerBriefs in Computational Mechanics, DOI: 10.1007/978-3-642-25311-9_7,
© The Authors(s) 2012

Fig. 7.1 Tensile creep
curves for rigid PVC
quenched from 90°C (about
10°C above T$_g$) to 40°C and
aged at 40°C for a period of
4 years. The curves were
obtained at the different aging
times shown [6]

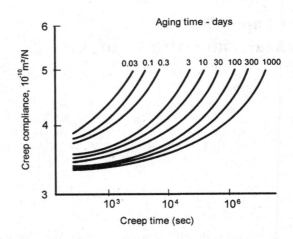

7.2 Viscoelastic Aging Formulation

A general relation for a linearly viscoelastic aging material can be written in the
history-dependent (uniaxial) form

$$\varepsilon(t) = \int_{\tau_0}^{t} D(t,\tau)\dot\sigma(\tau)d\tau = \frac{\sigma(t)}{E(t)} - \int_{\tau_0}^{t} \frac{\partial D(t,\tau)}{\partial\tau}\sigma(\tau)d\tau \qquad (7.1)$$

$D(t,\tau)$ is the response to a unit stress $\sigma(t) = H(t-\tau)$ applied at time τ, $D(t,\tau)$
has instantaneous and deferred components, whose separation is a matter of
convention to be discussed in each case. For nonaging viscoelastic materials it is

$$D(t,\tau) = D(t-\tau) = \frac{1}{E} + C(t-\tau) \qquad (7.2)$$

For aging materials two situations must be considered. The first one corresponds to *materials that reduce their compliance with age* because of the for-
mation of new bonds. In this case it is written

$$D(t,\tau) = \frac{1}{E(\tau)} + C(t,\tau) \qquad (7.3)$$

with $\partial E(\tau)/\partial\tau > 0$ and $\partial C(t,\tau)/\partial\tau < 0$. The instantaneous part in the creep
compliance is a function of the delayed time τ as the new bonds are initially
unstressed and only react to incremental loads. In absence of creep, $C(t,\tau) = 0$,
substitution of (7.3) into (7.1) yields

$$\varepsilon(t) = \int_{\tau_0}^{t} \frac{1}{E(\tau)}\dot\sigma(\tau)d\tau \qquad (7.4)$$

or, in rate form,

$$\dot{\varepsilon}(t) = \frac{\dot{\sigma}(t)}{E(t)} \tag{7.5}$$

For *materials whose stiffness degrades with age* because of the rupture of existing links, the instantaneous part is a function of the current time t

$$D(t, \tau) = \frac{1}{E(t)} + C(t, \tau) \tag{7.6}$$

with $\partial E(t)/\partial t < 0$ and $\partial C(t, \tau)/\partial t > 0$. For $C(t, \tau) = 0$,

$$\varepsilon(t) = \frac{\sigma(t)}{E(t)} \tag{7.7}$$

or, in rate form

$$\dot{\varepsilon}(t) = \frac{\dot{\sigma}(t)}{E(t)} - \frac{\sigma(t)\dot{E}(t)}{[E(t)]^2} \tag{7.8}$$

The concepts above may be extended to 3-D situations. For the case of isotropic materials, we have

$$\varepsilon_{ij}(t) = \int_{\tau_0}^{t} D_G(t, \tau)\dot{s}_{ij}(\tau)d\tau + \delta_{ij} \int_{\tau_0}^{t} D_K(t, \tau)\dot{\sigma}_0(\tau)d\tau \tag{7.9}$$

Where $D_G(t, \tau)$ and $D_K(t, \tau)$ are shear and volumetric creep functions respectively. For an anisotropic model for plates and shells case we have [7]

$$\varepsilon_i(t) = \int_0^t D_{ij}(t, \tau)\frac{\partial \sigma_j(\tau)}{\partial \tau}d\tau \quad (i, j = 1, 2, \ldots, 5) \tag{7.10}$$

where the composite is modelled as linear viscoelastic, with orthotropic symmetry. Then $\varepsilon_i(t)$ are the components of the strain vector $\{\varepsilon\} = \{\varepsilon_{11}, \varepsilon_{22}, 2\varepsilon_{12}, 2\varepsilon_{13}, 2\varepsilon_{23}\}$ and $\sigma_j(t)$ are the components of the stress vector $\{\sigma\} = \{\sigma_{11}, \sigma_{22}, \sigma_{12}, \sigma_{13}, \sigma_{23}\}$, at time t. Components ε_{33} and σ_{33} are not considered. $D_{ij}(t, \tau)$ are the creep functions corresponding to components ε_i and σ_j.

7.3 Rheological Models with Time Variable Parameters

Rheological models with time-dependent (or age-dependent) parameters have been proposed to represent the behavior of aging materials and are useful as they provide a physically intuitive representation.

Elastic model: The stress–strain relation for a spring with variable modulus $E(t)$ depends on the stress history. In the case of materials that harden with age, this relation has to be written in incremental or rate form

$$\dot{\sigma}(t) = E(t)\dot{\varepsilon}(t) \qquad (7.11)$$

which takes into account the physical processes involved in real materials (for example, the gradual solidification of cement paste or the gradual building of new links in polymer chains). It is important to notice that (7.11) and not $\dot{\sigma}_E(t) = E(t)\dot{\varepsilon}(t) + \dot{E}(t)\varepsilon(t)$ is to be employed. If the latter is used, we will have, for a constant stress test $\dot{\sigma}_E(t) = 0$ and therefore, $\dot{E}(t)\varepsilon(t) = -E(t)\dot{\varepsilon}(t)$. Thus, in a material that hardens with time $\left(E(t) > 0, \dot{E}(t) > 0\right)$ strains would decrease with time $(\dot{\varepsilon}(t) < 0)$; this result is not physically correct. From (7.11) we obtain the expressions for total stress and strain

$$\sigma(t) = \int\limits_{\tau_0}^{t} E(\tau)\dot{\varepsilon}(\tau)d\tau$$

$$\varepsilon(t) = \int\limits_{\tau_0}^{t} \frac{1}{E(\tau)}\dot{\sigma}(\tau)d\tau \qquad (7.12)$$

In the case of materials that softens with age (e.g., through a deterioration process) we have

$$\sigma(t) = E(t)\varepsilon(t) \qquad (7.13)$$

Kelvin model hardening case: adding equations of dashpot and spring we have

$$\sigma(t) = \eta(t)\dot{\varepsilon}(t) + \int\limits_{\tau_0}^{t} E(\tau)\dot{\varepsilon}(\tau)d\tau \qquad (7.14)$$

and differentiating (7.14) with relation to t

$$\dot{\sigma}(t) = \dot{\varepsilon}(t)[E(t) + \dot{\eta}(t)] + \eta(t)\ddot{\varepsilon}(t) \qquad (7.15)$$

With the additional constitutive assumption $E(t) + \dot{\eta}(t) = \eta(t)/\theta, \theta = const$, we obtain the differential equation

$$\ddot{\varepsilon} + \frac{\dot{\varepsilon}}{\theta} = \frac{\dot{\sigma}}{\eta(t)} \qquad (7.16)$$

which determines the creep function

$$D(t, \tau) = \frac{\theta}{\eta(\tau)}[1 - e^{-(t-\tau)/\theta}] \qquad (7.17)$$

that has the form $C(t, \tau) = \phi(\tau)f(t - \tau)$.

Kelvin model softening case: here we have

$$\sigma(t) = \eta(t)\dot{\varepsilon}(t) + E(t)\varepsilon(t) \tag{7.18}$$

With the additional constitutive assumption $E(t)/\eta(t) = 1/\theta = const$, the following relation is obtained

$$\varepsilon(t) = \int_0^t \frac{\sigma(\tau)}{\eta(\tau)} e^{-(t-\tau)/\theta} d\tau \tag{7.19}$$

This equation corresponds to the Arutyunyan model [1] used to represent the behavior of aging concrete.

7.4 Representation by Means of State Variables

We consider the viscoelastic relation in the integral form (7.1). As indicated above, we must work with strain rates. Thus, applying Leibnitz formula we have

$$\dot{\varepsilon}(t) = \int_{\tau_o}^t \frac{\partial}{\partial t}[D(t,\tau)\dot{\sigma}(\tau)]d\tau + D(t,t)\dot{\sigma}(t) \tag{7.20}$$

or

$$\dot{\varepsilon}(t) = \frac{\dot{\sigma}(t)}{E(t)} + \int_{\tau_o}^t \dot{D}(t,\tau)\dot{\sigma}(\tau)d\tau \tag{7.21}$$

where the relation $D(t,t) = 1/E(t)$ has been used.

As seen before, to get the state variables representation we must develop the creep function in Dirichlet-Prony series that can be chosen to give rise to Kelvin chains or Maxwell chains. For the case of an aging Kelvin chain we can define an exponential series expansion of the creep function

$$\dot{D}(t,\tau) = \sum_{i=1}^n \frac{1}{\eta_i(\tau)} e^{-(t-\tau)/\theta_i} \tag{7.22}$$

corresponding to a series of n Kelvin models with time dependent parameters. Defining state variables in the form

$$q_i(t) = \int_{\tau_o}^t \frac{\dot{\sigma}(\tau)}{\eta_i(\tau)} e^{-(t-\tau)/\theta_i} d\tau \tag{7.23}$$

(7.21) can be written

$$\dot{\varepsilon}(t) = \frac{\dot{\sigma}(t)}{E(t)} + \sum_{i=1}^{n} q_i(t) \tag{7.24}$$

For the isotropic 3-D case, we have the expressions

$$\dot{\boldsymbol{\varepsilon}}(t) = \frac{\dot{\boldsymbol{\sigma}}^o(t)}{3K(t)} + \sum_{i=1}^{n} \mathbf{q}_i^0(t) + \frac{\dot{\mathbf{s}}(t)}{2G(t)} + \sum_{i=1}^{m} \mathbf{q}_i^s(t) \tag{7.25}$$

with

$$\mathbf{q}_j^o(t) = \int_{\tau_o}^{t} \frac{1}{\eta_j^o(\tau)} e^{-(t-\tau)(t-\tau)/\theta_j^o} \dot{\boldsymbol{\sigma}}^o(\tau) d\tau$$

$$\mathbf{q}_k^s(t) = \int_{\tau_o}^{t} \frac{1}{\eta_k^s(\tau)} e^{-(t-\tau)/\theta_k^s} \dot{\mathbf{s}}(\tau) d\tau \tag{7.26}$$

for the hardening case and

$$\boldsymbol{\varepsilon}(t) = \frac{\boldsymbol{\sigma}^o(t)}{3K(t)} + \sum_{i=1}^{n} \mathbf{q}_i^0(t) + \frac{\mathbf{s}(t)}{2G(t)} + \sum_{i=1}^{m} \mathbf{q}_i^s(t) \tag{7.27}$$

with

$$\mathbf{q}_j^o(t) = \int_{\tau_o}^{t} \frac{1}{\eta_j^o(\tau)} e^{-(t-\tau)/\theta_j^o} \boldsymbol{\sigma}^o(\tau) d\tau$$

$$\mathbf{q}_k^s(t) = \int_{\tau_o}^{t} \frac{1}{\eta_k^s(\tau)} e^{-(t-\tau)/\theta_k^s} \mathbf{s}(\tau) d\tau \tag{7.28}$$

for the softening case. In these equations, $\boldsymbol{\sigma}^o$ and $\mathbf{s}$ are the hydrostatic and deviator stress tensors, respectively. Integration in time can be performed as indicated in Chap. 3.

7.5 Aging and the Time Shifting Procedure

The time shifting procedure developed to represent temperature effects and addressed in Chap. 6 has been extended by Struik [8] and other authors [3, 4, 9] to the case of physical aging. A very complete study by Hutchinson [6] describes the phenomenon of physical aging of polymers, the thermodynamic and micro-structural theories proposed for this phenomenon and the use of time shifting procedure to model its effect on the viscoelastic properties of the material.

Still, an essential difference between temperature and age effects has to be recognised. Age t and time τ in (7.1) are variables with particular characteristics: for example, $D(t, \tau) = 0$ for any $t < \tau$. We can make experimental tests with constant temperature, but cannot make tests with constant age, because age changes with time along the test. To avoid this problem, Struik [8] recommends to perform "momentarily" tests, with short duration, assuming that age is approximately constant in each of them. This is an interesting device, but one that does not provide complete information on the behavior of the aging material. Up to the present time, there is no single accepted method for reducing this short-term data, and different schemes can lead to significantly different results in both short and long-term predictions [3]. As in the case of temperature vertical shift is needed in addition to horizontal shift in order to represent adequately the mechanical behavior.

References

1. N. Kh. Arutyunyan, *Some problems in the theory of creep* (Pergamon Press, Oxford, 1966)
2. Z.P. Bazant, Theory of creep and shrinkage in concrete structures: a précis of recent developments. in *Mechanics Today*, ed. by S. Nemat-Nasser Pergamon Press, vol 2, pp. 1–93 (1975)
3. R.D. Bradshaw, L.C. Brinson, Physical aging in polymers and polymer composites: an analysis and method for time-aging time superposition. Polym. Eng. Sci. **37**(1), 31–44 (1997)
4. L.C. Brinson, T.S. Gates, Effects of physical aging on long term creep of polymers and polymer matrix composites. Int. J. Solids Struct. **32**(6–7), 827–846 (1995)
5. G.J. Creus, *Viscoelasticity-Basic Theory and Application to Concrete Structures*, (Springer, Berlin, 1986)
6. J.M. Hutchinson, Physical aging of polymers. Prog. Polym. Sci. **20**, 703–760 (1995)
7. B.F. Oliveira, G. Creus, An analytical-numerical framework for the study of ageing in fibre reinforced polymer composites. Compos. Struct. **65**(3–4), 443–457 (2004)
8. L.C.E. Struik, *Physical Aging in Amorphous Polymers and Other Materials* (Elsevier Scientific Publishing Co., New York, 1978)
9. J.L. Sullivan, Creep and physical aging of composites. Compos. Sci. Technol.**39**(3), 207–232 (1990)

Chapter 8
Nonlinear Viscoelasticity

Viscoelastic behavior may show physical and/or geometrical nonlinearity. Physical nonlinearity corresponds to situations in which the linear behavior described in Chap. 1 (Sect. 1.3.2) is not observed, even in small strain situations. Geometrical nonlinearity corresponds to situations of large deformations (large displacements and/or large strain). Both effects can appear combined in some problems (e.g. polymers, biomechanics). Alternative nonlinear or quasi-linear single integral representations have been proposed, some of which are described in Sect. 8.2. In Sect. 8.3, a nonlinear state variables formulation proposed by Simo is described. The situation involving large displacements associated with small strains that is particularly important in the analyses of materials and structures is addressed in detail in Chap. 9.

8.1 Schapery Single Integral Non-Linear Viscoelasticity

The Schapery single integral constitutive equation of non-linear viscoelasticity was derived from fundamental principles utilizing the concepts of irreversible thermodynamics [11] in a small strain context. For isothermal condition and uniaxial stress, the creep constitutive equation proposed by Schapery can be written as

$$\varepsilon(t) = g_0 D(0)\sigma(t) + g_1 \int_0^t C(\psi - \psi') \frac{\partial}{\partial \tau}(g_2 \sigma(\tau)) d\tau \tag{8.1}$$

where $D(0)$ and $C(\psi)$ are the instantaneous and transient components of the creep compliance in linear viscoelasticity, respectively. The arguments ψ and ψ' are

S. P. C. Marques and G. J. Creus, *Computational Viscoelasticity*,
SpringerBriefs in Computational Mechanics, DOI: 10.1007/978-3-642-25311-9_8,

reduced times that take into account simultaneously, temperature and stress effects and are given by

$$\psi = \int_0^t \frac{ds}{a(\theta)b(\sigma)} \text{ and } \psi' = \int_0^\tau \frac{ds}{a(\theta)b(\sigma)} \tag{8.2}$$

in which $a(\theta)$ is the temperature shift factor used for thermorheologically simple materials (see Chap. 6) whereas $b(\sigma)$ is the stress shift factor. Then, $D(\psi - \psi') = D(0) + C(\psi - \psi')$ is the creep compliance adjusted for stress and temperature variations.

The material properties g_0, g_1 and g_2 are nonlinear functions of stress. $g_0(\sigma)$ is related to the nonlinear instantaneous compliance. The transient creep parameter $g_1(\sigma)$ measures the nonlinear effect on creep and g_2 accounts for the load rate effect. When the applied stress is small, $g_0 = g_1 = g_2 = b(\sigma) = 1$ and (8.1) is reduced to the integral representation of linear viscoelastic behavior. Notice that when the mentioned stress functions are dependent on temperature, the material behavior is thermorheologically complex.

When strain is the independent variable, the corresponding relaxation constitutive equation for constant temperature is given by

$$\sigma(t) = h_\infty E(\infty)\varepsilon(t) + h_1 \int_0^t R(\zeta - \zeta')\frac{\partial}{\partial \tau}(h_2\varepsilon(\tau))d\tau \tag{8.3}$$

where h_∞, h_1 and h_2 are material nonlinear functions of the strain. $E(\infty)$ and $R(\zeta)$ indicate the asymptotic modulus at constant strain$(t \to \infty)$ and the transient component of the relaxation function, in linear viscoelasticity, respectively. The variables ζ and ζ' stand for reduced times defined by

$$\zeta = \int_0^t \frac{ds}{a(\theta)c(\varepsilon)} \qquad \zeta' = \int_0^\tau \frac{ds}{a(\theta)c(\varepsilon)} \tag{8.4}$$

where $c(\varepsilon)$ is the strain shift factor. Hence, considering the above definitions, the relaxation function of the material, adjusted to take into account temperature and strain effects, can be expressed as $E(\zeta - \zeta') = E(\infty) + R(\zeta - \zeta')$

As seen in (8.1) and (8.2), the Schapery formulation involves five material functions and a material constant to be determined experimentally. This number of experimental parameters justifies the power of the model to fit the nonlinear behavior of many viscoelastic materials. A detailed explanation of the complex procedure needed to obtain these parameters may be found in the book by Brinson and Brinson [1].

Several authors extended the Schapery model to 3D situations and implemented it into finite element procedures [6, 7, 9].

8.2 Nonlinear Viscoelasticity at Large Strains in Integral Form

8.2.1 General Constitutive Relation

In the case of large strain problems (geometrical nonlinearity) appropriate measures for stress and strain and their functions have to be used (see Appendix B).

Constitutive relations in a large deformation context need to be objective, that is, independent of the presence of large displacements and rotations.

We begin with

$$\boldsymbol{\sigma}(t) = \mathcal{G}_{\tau=0}^{t} (\mathbf{F}) \tag{8.5}$$

(compare it with the small strain version (1.6)) and look for the conditions to make it objective. To do this, rigid body motions, characterized by translations $\mathbf{d}(\tau)$ and rotations $\mathbf{Q}(\tau)$, $0 \leq \tau \leq t$, are superposed on the body motion $\mathbf{x}(\tau) = \chi(\mathbf{X}, \tau)$. For this case, the particle position and deformation gradient are given by $\mathbf{x}^*(\tau) = \mathbf{Q}(\tau)\mathbf{x}(\tau) + \mathbf{d}(\tau)$ and $\mathbf{F}^*(\tau) = \mathbf{Q}(\tau)\mathbf{F}$, respectively. When the material is subjected to a rotation $\mathbf{Q}$ ($\mathbf{Q}$ is an orthogonal tensor, $\mathbf{Q}\mathbf{Q}^{\mathbf{T}} = \mathbf{I}$), the stress $\boldsymbol{\sigma}$ transforms as a second order tensor while the deformation gradient $\mathbf{F}$ transforms as a vector. Then we have

$$\mathbf{Q}\,\boldsymbol{\sigma}\mathbf{Q}^{T} = \mathcal{G}_{\tau=0}^{t}(\mathbf{QF}) = \mathcal{G}_{\tau=0}^{t}(\mathbf{QRU}) \tag{8.6}$$

To find a *necessary* condition we choose $\mathbf{Q} = \mathbf{R}^{T} = \mathbf{R}^{-1}$ and substitute it into (8.6) obtaining

$$\boldsymbol{\sigma} = \mathcal{G}_{\tau=0}^{t}(\mathbf{F}) = \mathbf{R}\mathcal{G}_{\tau=0}^{t}(\mathbf{U})\,\mathbf{R}^{T} \tag{8.7}$$

and

$$\mathbf{F}^{T}\boldsymbol{\sigma}\mathbf{F} = \mathbf{U}F_{\tau=0}^{t}(\mathbf{C})\,\mathbf{U} = \tilde{\mathcal{G}}_{\tau=0}^{t}(\mathbf{C}) \tag{8.8}$$

because $\mathbf{F} = \mathbf{RU}$ and $\mathbf{C} = \mathbf{F}^{T}\mathbf{F} = \mathbf{U}^2$ (see Appendix B). Thus

$$\boldsymbol{\sigma}(t) = \mathbf{F}(t)\tilde{\mathcal{G}}[\mathbf{C}(t-\tau)]_{\tau=0}^{t}\,\mathbf{F}^{T}(t) \tag{8.9}$$

Using the relation between the Cauchy stress tensor $\boldsymbol{\sigma}$ and the Second Piola–Kirchhoff stress tensor $\mathbf{S}$ (see Appendix B) in (8.9), we obtain

$$\mathbf{S}(t) = J(t)\tilde{\mathcal{G}}_{\tau=0}^{\tau=t}\{\mathbf{C}(\tau)\} \tag{8.10}$$

The next problem is to describe the nonlinear functional.

8.2.2 Multiple Integral Representations

A formulation to describe nonlinear viscoelastic functionals was given by Volterra using an earlier representation developed by Frechet in the early 1900's. This formulation was forgotten until the procedure was generalized to three dimensions by Rivlin and Green.

Assuming that the response functional obeys the continuity condition required by the *weak principle of fading memory* [14], Green and Rivlin [5] derived an approximate integral constitutive relation. Considering that the functional $\tilde{G}$ in (8.9) is continuous in $\mathbf{C}(\tau)$, $0 \leq \tau \leq t$, they used the Stone-Weierstrass theorem and the Fourier expansion of polynomials by integrals to derive the multiple integral representation

$$
\boldsymbol{\sigma}(t) = \mathbf{F}(t) \left[\int_0^t K_1(t - \tau_1) \mathbf{C}(\tau_1) d\tau_1 \right.
$$
$$
\left. + \int_0^t \int_0^t \mathbf{K}_2(t - \tau_1, t - \tau_2) \mathbf{C}(\tau_1) \mathbf{C}(\tau_2) d\tau_1 d\tau_2 + \ldots \right] \mathbf{F}^T(t) \tag{8.11}
$$

where the memory kernels $\mathbf{K}_k$, $k = 1, 2, \ldots, n$, are positive, continuous and monotonically decreasing tensor-valued functions of time. The number of terms required in (8.11) to obtain an adequate approximation depends on the characteristics of the strain history. Findley et al. [3] describes the procedure both theoretically and experimentally. Because of the difficulty to evaluate experimentally a large number of functions and because of stability problems, expansions are limited to the third order. The experimental determination of parameters [15] is difficult and the numerical computations are time consuming. Thus, this representation is seldom applied to the solution of practical problems [12].

Another multiple integral representation has been proposed by Coleman and Noll [14]

8.2.3 Pipkin–Rogers Model

The nonlinear viscoelastic constitutive theory presented by Pipkin and Rogers [10] is based on the analysis of the response of the material to step strain histories. According to this model, the functional in (8.9) can be expanded in a series whose first term provides the best approximation to measured mechanical behavior using single step strain histories. This leading term is given by

$$
\boldsymbol{\sigma}(t) = \mathbf{F}(t) \left\{ \mathbf{K}[\mathbf{C}(t), 0] + \int_0^t \frac{\partial}{\partial \tau} \mathbf{K}[\mathbf{C}(\tau), t - \tau] d\tau \right\} \mathbf{F}^T(t) \tag{8.12}
$$

$\mathbf{K}(\mathbf{C}, t)$ is the strain dependent relaxation tensor induced by a single step strain history and has the form $\mathbf{K} = \Phi_0 \mathbf{I} + \Phi_1 \mathbf{C} + \Phi_2 \mathbf{C}^2$, where Φ_0, Φ_1 and Φ_2 are scalar functions of t and the invariants of $\mathbf{C}$ [2, 16].

8.2.4 Quasi-Linear Viscoelastic Model

When the constitutive tensor $\mathbf{K}(\mathbf{C}(\tau), t - \tau)$ appearing in (8.12) can be decomposed in the form

$$\mathbf{K}[\mathbf{C}, t - \tau] = \mathbf{K}_e[\mathbf{C}]F(t - \tau) \tag{8.13}$$

with $F(0) = 1$, the formulation is known as quasi-linear viscoelasticity. For this case, (8.12) can be expressed as

$$\sigma(t) = \mathbf{F}(t)\left\{ \mathbf{K}_e[\mathbf{C}(t)] + \int_0^t \mathbf{K}_e[\mathbf{C}(\tau)] \frac{\partial F(t - \tau)}{\partial (t - \tau)} d\tau \right\} \mathbf{F}^T(t) \tag{8.14}$$

This constitutive relation was proposed by Fung [4] and used for modelling the mechanical behavior of biological tissues. The terminology "quasi-linear viscoelasticity" is used because $\mathbf{K}[\mathbf{C}]$ can be thought of as a nonlinear measure of strain. The expression in braces in (8.14) is linear in this nonlinear strain measure.

8.3 Nonlinear Viscoelasticity at Large Strains Using State Variables

Simo and co-workers [13] developed a constitutive model for nonlinear viscoelasticity based on state variables. It is particularly addressed to materials like polymers and rubbers that behave as hyperelastic in short time loading situations and uses the concepts of deviatoric-volumetric split that is also convenient from the computational point of view. This model assumes that viscoelastic behavior is restricted to shear and that bulk strain is purely elastic. According to Simo the formulation has the following attractive features:

1. It uses the numerical implementation of incremental integration as described in Chap. 3.
2. It is a description of time-dependent behavior that contains hyperelasticity as a particular case.
3. It allows a separation of volume preserving and dilatational responses.

The development follows the pattern of linear viscoelasticity: the time-dependent behavior reduces to the corresponding hyperelastic behavior for very fast or very slow processes and the state variables formulation is an extension of the same formulation for small strain. Thus, we will describe first the hiperelastic formulation, then a state variable formulation for small strains recast in a slightly different way to allow for the introduction of the strain energy and finally the finite strain formulation.

8.3.1 Hyperelastic Formulation

A hyperelastic material is characterized by a *strain energy function*

$$\Psi = \Psi(\mathbf{F}) = \Psi(\mathbf{C}) = \Psi(\mathbf{E}) \geq 0 \text{ with } \Psi(\mathbf{F} = \mathbf{I}) = 0 \tag{8.15}$$

The constitutive relation for a hyperelastic material is by definition

$$\mathbf{S} = 2\frac{\partial \Psi(\mathbf{C})}{\partial \mathbf{C}} = \frac{\partial \Psi(\mathbf{E})}{\partial \mathbf{E}} \tag{8.16}$$

Volumetric-shear split

Some materials (e.g., polymers) behave quite differently in bulk and in shear. Then it may be convenient to split the deformation into a volumetric part and an isochoric part, as it is done in the small strain case (Sect. 4.2.1). This split has advantages also from the computational point of view. Thus, we use the multiplicative decomposition [8]

$$\bar{\mathbf{F}} = J^{-1/3}\mathbf{F} \tag{8.17}$$

where $J = \det \mathbf{F}$ is related to the volume changes, i.e., to the dilatational part of $\mathbf{F}$, and $\bar{\mathbf{F}}$ is associated to the volume-preserving or isochoric part of $\mathbf{F}$. Notice that, from (8.17), $\bar{J} = \det \bar{\mathbf{F}} = 1$.

Introducing (8.17) into the definition of the *right Cauchy-Green tensor* $\mathbf{C} = \mathbf{F}^T\mathbf{F}$, we have

$$\mathbf{C} = J^{2/3}\bar{\mathbf{C}} \tag{8.18}$$

being $\bar{\mathbf{C}} = \bar{\mathbf{F}}^T\bar{\mathbf{F}}$ the isochoric part of $\mathbf{C}$. Similarly, the strain energy is divided into volumetric and isochoric parts

$$\Psi(\mathbf{C}) = \Psi_{vol}(J) + \Psi_{iso}(\bar{\mathbf{C}}) \tag{8.19}$$

Thus,

$$\dot{\Psi} = \frac{\partial \Psi_{vol}(J)}{\partial J}\dot{J} + \frac{\partial \Psi_{iso}(\bar{\mathbf{C}})}{\partial \bar{\mathbf{C}}} : \dot{\bar{\mathbf{C}}} = p\dot{J} + \frac{1}{2}\bar{\mathbf{S}} : \dot{\bar{\mathbf{C}}} \tag{8.20}$$

with $p = \frac{\partial \Psi_{vol}(J)}{\partial J}$ and $\bar{\mathbf{S}} = 2\frac{\partial \Psi_{iso}(\bar{\mathbf{C}})}{\partial \bar{\mathbf{C}}}$

The Second Piola–Kirchhoff stress $\mathbf{S}$ *is also* divided into isochoric and volumetric parts

$$\mathbf{S} = 2\frac{\partial\Psi(\mathbf{C})}{\partial\mathbf{C}} = \mathbf{S}_{iso} + \mathbf{S}_{vol} \tag{8.21}$$

where

$$\mathbf{S}_{vol} = 2\frac{\partial\Psi_{vol}(J)}{\partial\mathbf{C}} = J\frac{\partial\Psi_{vol}(J)}{\partial J}\mathbf{C}^{-1} = Jp\mathbf{C}^{-1} \tag{8.22}$$

$$\mathbf{S}_{iso} = 2\frac{\partial\Psi_{iso}(\bar{\mathbf{C}})}{\partial\mathbf{C}} = J^{-2/3}DEV\bar{\mathbf{S}}$$

with $DEV\bar{\mathbf{S}} = \left(\mathbb{I} - \frac{1}{3}\mathbf{C}\otimes\mathbf{C}^{-1}\right):\bar{\mathbf{S}}$. In this relation, $\mathbb{I}$ is the fourth-order unit tensor and $\otimes$ denotes a tensor product.

Example 1: Derive the relations (8.22).

Applying the chain rule to (8.22)$_1$ and using the relation $\frac{\partial J}{\partial\mathbf{C}} = \frac{J}{2}\mathbf{C}^{-1}$, we have

$$\mathbf{S}_{vol} = 2\frac{\partial\Psi_{vol}(J)}{\partial\mathbf{C}} = 2\frac{\partial\Psi_{vol}(J)}{\partial J}\frac{\partial J}{\partial\mathbf{C}} = 2\frac{\partial\Psi_{vol}(J)}{\partial J}\frac{J}{2}\mathbf{C}^{-1} = Jp\mathbf{C}^{-1}$$

Now, we apply the chain rule to (8.22)$_2$ and use the definition of $\bar{\mathbf{S}}$ to obtain

$$\mathbf{S}_{iso} = 2\frac{\partial\Psi_{iso}(\bar{\mathbf{C}})}{\partial\mathbf{C}} = 2\frac{\partial\Psi_{iso}(\bar{\mathbf{C}})}{\partial\bar{\mathbf{C}}}\frac{\partial\bar{\mathbf{C}}}{\partial\mathbf{C}} = \bar{\mathbf{S}}:\frac{\partial\bar{\mathbf{C}}}{\partial\mathbf{C}}$$

where, from (8.18),

$$\frac{\partial\bar{\mathbf{C}}}{\partial\mathbf{C}} = \frac{\partial\left(J^{-2/3}\mathbf{C}\right)}{\partial\mathbf{C}} = J^{-2/3}\left(\mathbb{I} - \frac{2}{3}\frac{1}{J}\frac{\partial J}{\partial\mathbf{C}}\otimes\mathbf{C}\right) = J^{-2/3}\left(\mathbb{I} - \frac{1}{3}\mathbf{C}\otimes\mathbf{C}^{-1}\right)$$

and therefore, $\mathbf{S}_{iso} = J^{-2/3}DEV\bar{\mathbf{S}}$, where the *DEV* operator delivers the deviatoric part of the stress tensor.

8.3.2 Viscoelastic Small Strain Relations

Simo proposes a visco-hyperelastic model which is an extension of the linear viscoelastic formulation described here in Chap. 3, recast in a convenient format.

He begins with a state variables formulation similar to that introduced in Sect. 3.1 for the generalized Maxwell model, modified to make the extension to finite strains easier. First, a new internal variable $\hat{q}_i$ is introduced, so that

$$\hat{q}_i = E(\varepsilon - q_i) \tag{8.23}$$

where q_i $(i = 1,\ldots,n)$ are the state variables defined for the generalized Maxwell model in Chap. 3. Thus, (3.9) and (3.10) become

$$\dot{\hat{q}}_i + \frac{\hat{q}_i}{T_i} = \frac{\gamma_i}{T_i} E_0 \varepsilon$$

$$\sigma = E_0 \varepsilon - \sum_{i=1}^{n} \hat{q}_i \tag{8.24}$$

where $E_0 = E(0) = E_\infty + \sum_{i=1}^{n} E_i$ and $\gamma_i = E_i/E_0$.

In the small strain three-dimensional context the strain energy can be decomposed in isochoric and volumetric parts

$$\Psi^0(\varepsilon) = \Psi^0_{iso}(\mathbf{e}) + \Psi^0_{vol}(tr\varepsilon) \tag{8.25}$$

Considering that bulk deformation is elastic (8.24) may be written

$$\dot{\hat{\mathbf{q}}}_{\mathbf{i}} + \frac{\hat{\mathbf{q}}_{\mathbf{i}}}{T_i} = \frac{\gamma_i}{T_i} \frac{\partial \Psi^0(\mathbf{e})}{\partial \mathbf{e}} \tag{8.26}$$

$$\sigma = \frac{\partial \Psi^0(\varepsilon)}{\partial \varepsilon} - \sum_{i=1}^{n} \hat{\mathbf{q}}_{\mathbf{i}}$$

8.3.3 Formulation of the Nonlinear Viscoelastic Model

The generalization of $(8.24)_2$ to the finite deformation regime is

$$\mathbf{S}(t) = \mathbf{S}^0(t) - J^{-2/3} DEV\left[\sum_{i=1}^{n} \mathbf{Q}_i(t)\right] \tag{8.27}$$

where $\mathbf{S}^0(t)$ is given by (8.21) with Ψ being the total initial stored-energy function $\Psi^0 = \Psi^0_{iso} + \Psi^0_{vol}$ and $\mathbf{C}(t)$, being a function of time.

The growth law for the internal state variables is written, following (8.26)

$$\dot{\mathbf{Q}}_i(t) + \frac{1}{T_i} \mathbf{Q}_i(t) = \frac{\gamma_i}{T_i} DEV\left(2\frac{\partial \Psi^0_{iso}(\bar{\mathbf{C}}(t))}{\partial \bar{\mathbf{C}}}\right) \tag{8.28}$$

with $\mathbf{Q}_i(t \leq \tau_0) = \mathbf{0}$. Ψ^0_{iso} denotes the volume-preserving contribution to the stored-energy function.

The solution of the differential equation (8.28) has the integral representation

$$\mathbf{Q}_i(t) = \frac{\gamma_i}{T_i} \int_{-\infty}^{t} e^{-(t-\tau)/T_i} DEV\left(2\frac{\partial \Psi_{iso}^0(\bar{\mathbf{C}}(\tau))}{\partial \bar{\mathbf{C}}}\right) d\tau \qquad (8.29)$$

These expressions are formally similar to the corresponding expressions in Chap. 3. The recurrence formula for determination of $\mathbf{Q}_i$ is also similar to that shown in Sect. 3.2.

Time integration algorithm

At time t_n we assume to know the displacement field $\mathbf{u}_n$, its dependent variables $\mathbf{F}_n = \mathbf{I} + \partial \mathbf{u}_n/\partial \mathbf{X}$, $J_n = \det \mathbf{F}_n$, $\mathbf{C}_n = \mathbf{F}_n^T \mathbf{F}_n$, $\bar{\mathbf{C}}_n = J^{-2/3} \mathbf{C}_n$ and the stress $\mathbf{S}_n$ satisfying the equilibrium conditions. We need to determine the updated values of these variables for a new displacement field $\mathbf{u}_{n+1}$, at time $t_{n+1} = t_n + \Delta t$, which is corrected iteratively until the balance equations are satisfied within the given tolerance. For this, we can use the following time integration algorithm:

(1) Given initial information at time t_n : $\bar{\mathbf{S}}_n^0$, $\mathbf{C}_n$ and $(\mathbf{Q}_i)_n$ with $i = 1,\ldots,m$;

(2) For a given trial solution $\mathbf{u}_{n+1}$ at time t_{n+1}, compute $\mathbf{F}_{n+1} = \mathbf{I} + \partial \mathbf{u}_{n+1}/\partial \mathbf{X}$, $J_{n+1} = \det \mathbf{F}_{n+1}$, $\mathbf{C}_{n+1} = \mathbf{F}_{n+1}^T \mathbf{F}_{n+1}$, $\bar{\mathbf{C}}_{n+1} = J_{n+1}^{-2/3} \mathbf{C}_{n+1}$;

(3) Evaluate $\bar{\mathbf{S}}_{n+1}^0 = \left[2\frac{\partial \Psi^0(\bar{\mathbf{C}})}{\partial \mathbf{C}}\right]_{n+1}$, $(\mathbf{S}_{iso}^0)_{n+1} = J_{n+1}^{-2/3} DEV\bar{\mathbf{S}}_{n+1}^0$,

$p_{n+1}^0 = \left(\frac{\partial \Psi_{vol}^0}{\partial J}\right)_{n+1}$, $(\mathbf{S}_{vol}^0)_{n+1} = J_{n+1}p_{n+1}^0 \mathbf{C}_{n+1}^{-1}$, $\mathbf{S}_{n+1}^0 = (\mathbf{S}_{iso}^0)_{n+1} + (\mathbf{S}_{vol}^0)_{n+1}$;

(4) Update state variables and stresses: $(\mathbf{Q}_i)_{n+1} = e^{-\Delta t/T_i}(\mathbf{Q}_i)_n + \frac{\gamma_i}{2}\left(DEV\bar{\mathbf{S}}_{n+1}^0 + DEV\bar{\mathbf{S}}_n^0\right)\left(1 - e^{-\Delta t/T_i}\right)$ and $\mathbf{S}_{n+1} = \mathbf{S}_{n+1}^0 - J_{n+1}^{-2/3}DEV\left[\sum_{i=1}^{m}(\mathbf{Q}_i)_{n+1}\right]$

An alternative time integration algorithm can be found in Holtzapfel [8].

References

1. H.F. Brinson, L.C. Brinson, *Polymer Engineering Science and viscoelasticity: an Introduction* (Springer, New York, 2008)
2. C.S. Drapaca, S. Sivaloganathan, G. Tenti, Nonlinear constitutive laws in viscoelasticity. Math. Mech. Solids **12**, 475–501 (2007)
3. W.N. Findley, J.S. Lai, K. Onaran, *Creep and Relaxation of Nonlinear Viscoelastic Materials* (Dover Publications Inc, New York, 1989)
4. Y.C. Fung, *Biomechanics Mechanical Properties of Living Tissues* (Springer, New York, 1993)
5. A.E. Green, R.S. Rivlin, The mechanics of non-linear materials with memory. Arch. Ration. Mech. Anal. **1**, 1–21 (1957)
6. R.M. Haj-Ali, A.H. Muliana, Numerical finite element formulation of the Schapery nonlinear viscoelastic material model. Int. J. Numer. Meth. Engng **59**(1):25–45 (2004)
7. M. Henriksen, Nonlinear viscoelastic stress analysis—a finite element approach. Comput. Struct. **18**(1), 133–139 (1984)

8. G.A. Holtzapfel, *Nonlinear Solid Mechanics* (Wiley, West Sussex, 2004)
9. J. Lai, A. Bakker, 3-D Schapery representation for non-linear viscoelasticity and finite element implementation. Comput. Mech **18**, 182–191 (1996)
10. A.C. Pipkin, T.G. Rogers, Representation for viscoelastic behaviour. J. Mech. Phys. Solids **16**, 59–72 (1968)
11. R.A. Schapery, On the characterization of non-linear viscoelastic materials. Polym. Eng. Sci. **9**(4), 295–310 (1969)
12. R.A. Schapery, Nonlinear viscoelastic solids. Int. J. Solids. Struct. **37**, 359–366 (2000)
13. J.C. Simo, T.J.R. Hughes, *Computational Inelasticity* (Springer, New York, 1998)
14. C. Truesdell, W. Noll, *The Non-Linear Field Theories of Mechanics* (Springer, Berlin, 2004)
15. I.M. Ward, E.T. Onat, Nonlinear mechanical behaviour of oriented polypropylene. J. Mech. Phys. Solids **11**(4), 217–229 (1963)
16. A. Wineman, Nonlinear Viscoelastic Solids: a review. Math. Mech. Solids **14**, 300–366 (2009)

Chapter 9
Viscoelastic Finite Element Formulation

The finite element method is the most popular numerical procedure for the analysis of solids and structures, including those with time dependent properties. In this chapter, we present an incremental viscoelastic finite element formulation for problems with geometrical nonlinearity characterized by large displacements and rotations with small strains. The formulation is based on a total Lagrangian kinematic description. We begin with a brief presentation on the principle of virtual displacements for geometrically nonlinear problems. Procedures used for the computational implementation of the nonlinear viscoelastic model are also presented. We assume that the reader has a basic knowledge of the finite element method and of nonlinear continuum mechanics.

9.1 Principle of Virtual Displacements

Let us consider the motion of a body with arbitrary large displacements and rotations. Figure 9.1 shows the body configurations C^0, C^t and $C^{t+\Delta t}$ at instants τ_0, t and $t + \Delta t$, respectively, and the fixed coordinate system used as reference for the static and kinematic variable. We are interested in evaluating the body equilibrium in a finite sequence of configurations corresponding to times t_1, t_2,...,t_n within the analysis time range. As strategy used in this evaluation, we assume that the variable fields in the configuration $C^{t+\Delta t}$ can be completely determined if the solutions at times $\tau \leq t$ are already known.

The equilibrium condition of the body at time $t + \Delta t$ can be established by the principle of virtual displacements, as follows

$$\int_{\Omega^{t+\Delta t}} \sigma^{t+\Delta t} : \delta\varepsilon d\Omega^{t+\Delta t} = \int_{\Omega^{t+\Delta t}} \left(\mathbf{b}^{t+\Delta t}\right)^T \delta\mathbf{u} d\Omega^{t+\Delta t} + \int_{\Gamma^{t+\Delta t}} \left(\mathbf{t}^{t+\Delta t}\right)^T \delta\mathbf{u} d\Gamma^{t+\Delta t} \quad (9.1)$$

S. P. C. Marques and G. J. Creus, *Computational Viscoelasticity*,
SpringerBriefs in Computational Mechanics, DOI: 10.1007/978-3-642-25311-9_9,
© The Author(s) 2012

Fig. 9.1 Body
configurations and coordinate
systems

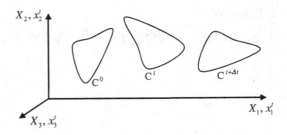

where the first member is the virtual work of the internal forces, whereas the
second member represents the virtual work of the external forces, i.e., body forces
b and surface forces **t**. In (9.1), $\delta\boldsymbol{\varepsilon}$ represents a variation in the infinitesimal strains
associated to the virtual increment $\delta\mathbf{u}$ in the displacement $\mathbf{u}^{t+\Delta t}$. The superscripts
stand for the instant of time at which the quantities are determined. The integrals
appearing in (9.1) are evaluated over the domain $\Omega^{t+\Delta t}$ and its boundary $\Gamma^{t+\Delta t}$
corresponding to configuration $C^{t+\Delta t}$.

For geometrically nonlinear analyses different definitions of stress and strain
tensors are used depending on the characteristics of the problem. In the present
development we use the *second Piola–Kirchhoff stress tensor* **S** and the
Green–Lagrange strain tensor **E** (see Appendix B) that are energetically conju-
gated (1), i. e.,

$$\int_{\Omega^{t+\Delta t}} \boldsymbol{\sigma}^{t+\Delta t} : \delta\boldsymbol{\varepsilon} d\Omega^{t+\Delta t} = \int_{\Omega^0} \mathbf{S}_0^{t+\Delta t} : \delta\mathbf{E}_0^{t+\Delta t} d\Omega^0 \tag{9.2}$$

where the index 0 is used to indicate that the quantities are referred to the initial
configuration C^0. Substituting (9.2) into (9.1), we have

$$\int_{\Omega^0} \mathbf{S}_0^{t+\Delta t} : \delta\mathbf{E}_0^{t+\Delta t} d\Omega^0 = \int_{\Omega^{t+\Delta t}} \left(\mathbf{b}^{t+\Delta t}\right)^T \delta\mathbf{u} d\Omega^{t+\Delta t} + \int_{\Gamma^{t+\Delta t}} \left(\mathbf{t}^{t+\Delta t}\right)^T \delta\mathbf{u} d\Gamma^{t+\Delta t} \tag{9.3}$$

As the Second Piola–Kirchhoff stress tensor and the Green–Lagrange strain
tensor are independent from the rigid body rotations, we may write

$$\mathbf{S}_0^{t+\Delta t} = \mathbf{S}_0^t + \Delta\mathbf{S}_0 \tag{9.4}$$

$$\mathbf{E}_0^{t+\Delta t} = \mathbf{E}_0^t + \Delta\mathbf{E}_0 = \mathbf{E}_0^t + \Delta\mathbf{e}_0 + \Delta\boldsymbol{\eta}_0$$

where $\Delta\mathbf{S}_0$ and $\Delta\mathbf{E}_0$ are the increments of the stress and strain measures between
t and $t + \Delta t$, respectively. In (9.4), $\Delta\mathbf{E}_0$ is decomposed in a linear part $\Delta\mathbf{e}_0$ and
nonlinear part $\Delta\boldsymbol{\eta}_0$, which in index notation are defined by

$$\Delta e_{0ij} = \frac{1}{2}\left(\frac{\partial\Delta u_i}{\partial X_j} + \frac{\partial\Delta u_j}{\partial X_i} + \frac{\partial u_k^t}{\partial X_i}\frac{\partial\Delta u_k}{\partial X_j} + \frac{\partial u_k^t}{\partial X_j}\frac{\partial\Delta u_k}{\partial X_i}\right) \tag{9.5}$$

$$\Delta\eta_{0ij} = \frac{1}{2}\frac{\partial\Delta u_k}{\partial X_i}\frac{\partial\Delta u_k}{\partial X_j}$$

where $\Delta\mathbf{u} = \mathbf{u}^{t+\Delta t} - \mathbf{u}^t$ is the displacement increment vector and $\mathbf{X} = (X_1,X_2,X_3)$ is the particle position in the initial configuration.

Substituting (9.4) into (9.3) and considering that the external loading is independent of the deformation, we obtain the total Lagrangian formulation of the incremental principle of virtual displacements as

$$\int_{\Omega^0}\Delta\mathbf{S}_0 : \delta(\Delta\mathbf{E}_0)d\Omega^0 + \int_{\Omega^0}\mathbf{S}_0^t : \delta(\Delta\boldsymbol{\eta}_0)d\Omega^0 = \int_{\Omega^0}\left(\mathbf{b}_0^{t+\Delta t}\right)^T\delta\mathbf{u}d\Omega^0$$
$$+ \int_{\Gamma^0}\left(\mathbf{t}_0^{t+\Delta t}\right)^T\delta\mathbf{u}d\Gamma^0 - \int_{\Omega^0}\mathbf{S}_0^t : \delta(\Delta\mathbf{e}_0)d\Omega^0 \tag{9.6}$$

being $\mathbf{b}_0^{t+\Delta t}$ and $\mathbf{t}_0^{t+\Delta t}$ the body and surface forces at time $t + \Delta t$, respectively, measured with respect to the initial configuration.

9.2 Linearization of the Principle of Virtual Displacements

We consider a viscoelastic body subjected to both mechanical and hygrothermal loads. For this case the total increment of the Green–Lagrange strain tensor at time interval $[t,t + \Delta t]$ is given by

$$\Delta\mathbf{E}_0 = \Delta\mathbf{E}_0^e + \Delta\mathbf{E}_0^V + \Delta\mathbf{E}_0^{HT} \tag{9.7}$$

where the superscripts e, V and HT are used to indicate the elastic, viscoelastic and hygrothermal contributions, respectively. Neglecting the effect of the nonlinear part $\Delta\boldsymbol{\eta}_0$ as an approximation to obtain the increment of the second Piola–Kirchhoff stress tensor, we may write

$$\Delta\mathbf{S}_0 \cong \mathbf{C}^e(\Delta\mathbf{e}_0 - \Delta\mathbf{e}_0^V - \Delta\mathbf{e}_0^{HT}) \tag{9.8}$$

where $\mathbf{C}^e$ is the 4th order elastic stiffness tensor of the material. Thus, using (9.8), the incremental principle of virtual displacements (9.6) can be rewritten in the form

$$\int_{\Omega^0}\mathbf{C}^e\Delta\mathbf{e}_0 : \delta(\Delta\mathbf{e}_0)d\Omega^0 + \int_{\Omega^0}\mathbf{S}_0^t : \delta(\Delta\boldsymbol{\eta}_0)d\Omega^0 = \int_{\Omega^0}\left(\mathbf{b}_0^{t+\Delta t}\right)^T\delta\mathbf{u}d\Omega^0 + \int_{\Gamma^0}\left(\mathbf{t}_0^{t+\Delta t}\right)^T\delta\mathbf{u}d\Gamma^0$$
$$+ \int_{\Omega^0}\mathbf{C}^e\Delta\mathbf{e}_0^V : \delta(\Delta\mathbf{e}_0)d\Omega^0 + \int_{\Omega^0}\mathbf{C}^e\Delta\mathbf{e}_0^{HT} : \delta(\Delta\mathbf{e}_0)d\Omega^0 - \int_{\Omega^0}\mathbf{S}_0^t : \delta(\Delta\mathbf{e}_0)d\Omega^0 \tag{9.9}$$

Equation (9.9) is the linearized form of the incremental principle of virtual displacements which will be used to derive the nonlinear finite element formulation in Sect. 9.3. It is worth noticing that to obtain (9.9) the approximation $\delta(\Delta \mathbf{E}_0) \cong \delta(\Delta \mathbf{e}_0)$ was used.

From relations (9.5), we may show that

$$\delta\left(\Delta e_{0ij}\right) = \frac{1}{2}\left[\delta\left(\frac{\partial \Delta u_i}{\partial X_j}\right) + \delta\left(\frac{\partial \Delta u_j}{\partial X_i}\right) + \frac{\partial u_k^t}{\partial X_i}\delta\left(\frac{\partial \Delta u_k}{\partial X_j}\right) + \frac{\partial u_k^t}{\partial X_j}\delta\left(\frac{\partial \Delta u_k}{\partial X_i}\right)\right] \quad (9.10)$$

$$\delta\left(\Delta \eta_{0ij}\right) = \frac{1}{2}\left[\frac{\partial \Delta u_k}{\partial X_i}\delta\left(\frac{\partial \Delta u_k}{\partial X_j}\right) + \frac{\partial \Delta u_k}{\partial X_j}\delta\left(\frac{\partial \Delta u_k}{\partial X_i}\right)\right]$$

9.3 Nonlinear Viscoelastic Finite Element Formulation

In an incremental geometrically nonlinear analysis, the total displacements in the current configuration $\mathbf{u}^{t+\Delta t}$ are obtained by adding the displacement increments $\Delta \mathbf{u}$ to the point coordinates $\mathbf{x}^t$ corresponding to the last configuration

$$\mathbf{u}^{t+\Delta t} = \mathbf{x}^t + \Delta \mathbf{u} \quad (9.11)$$

This is why it is convenient to use the same interpolation functions for displacement and coordinates (or geometry). The same interpolation functions used in the linear isoparametric finite element formulation can be employed for the nonlinear approach. Thus, for the three-dimensional case, the coordinate vector $\mathbf{X} = [X_1 \ \ X_2 \ \ X_3]^T$ of a finite element, with N nodal points, are in general defined in the initial configuration as

$$\mathbf{X}^{(e)} = \mathbf{H}(\xi)\tilde{\mathbf{X}}^{(e)} \quad (9.12)$$

where $\tilde{\mathbf{X}}^{(e)} = \left[\tilde{\mathbf{X}}_1^{(e)T} \ \ \tilde{\mathbf{X}}_2^{(e)T} \ \ \ldots \ \ \tilde{\mathbf{X}}_N^{(e)T}\right]^T$ is the element nodal coordinate vector being $\tilde{\mathbf{X}}_k^{(e)} = \left[\tilde{X}_1^{(e)} \ \ \tilde{X}_2^{(e)} \ \ \tilde{X}_3^{(e)}\right]_k^T$ the coordinate vector of the element k-node. In (9.12), $\mathbf{H}(\xi)$ represents the interpolation function matrix which has the general form

$$\mathbf{H}(\xi) = [\mathbf{H}_1(\xi) \ \ \mathbf{H}_2(\xi) \ \ \ldots \ \ \mathbf{H}_N(\xi)] \quad (9.13)$$

with the diagonal submatrices

$$\mathbf{H}_k(\xi) = N_k(\xi)\begin{bmatrix} 1 & 0 & 0 \\ 0 & 1 & 0 \\ 0 & 0 & 1 \end{bmatrix} \quad (9.14)$$

where $N_k(\xi)$, $(k = 1,2,\ldots,N)$, indicate the element interpolation functions whose argument is the natural coordinates ξ.

Similarly, the element displacement vector $\mathbf{u}^{(e)} = \begin{bmatrix} u_1^{(e)} & u_2^{(e)} & u_3^{(e)} \end{bmatrix}^T$ is given by the approximation

$$\mathbf{u}^{(e)} = \mathbf{H}(\boldsymbol{\xi})\tilde{\mathbf{u}}^{(e)} \tag{9.15}$$

where $\tilde{\mathbf{u}}^{(e)} = \begin{bmatrix} \tilde{\mathbf{u}}_1^{(e)T} & \tilde{\mathbf{u}}_2^{(e)T} & \cdots & \tilde{\mathbf{u}}_N^{(e)T} \end{bmatrix}^T$ represents the element nodal displacement vector and its components $\tilde{\mathbf{u}}_k^{(e)} = \begin{bmatrix} \tilde{u}_1^{(e)} & \tilde{u}_2^{(e)} & \tilde{u}_3^{(e)} \end{bmatrix}_k^T$ are the nodal displacement of the element k-node.

To simplify the finite element equations we use in this section Voigt notation. Then, for the three-dimensional case, the second Piola–Kirchhoff stress vector is given by

$$\hat{\mathbf{S}} = \begin{bmatrix} S_{11} & S_{22} & S_{33} & S_{23} & S_{13} & S_{12} \end{bmatrix}^T \tag{9.16}$$

and the Green–Lagrange strain vector by

$$\hat{\mathbf{E}} = \begin{bmatrix} E_{11} & E_{22} & E_{33} & 2E_{23} & 2E_{13} & 2E_{12} \end{bmatrix}^T \tag{9.17}$$

In Voigt notation, the shear components in (9.17) are doubled to allow writing the internal virtual work per volume unit as $\hat{\mathbf{S}}^T \delta \hat{\mathbf{E}}$. Then, the equilibrium equation (9.9) is expressed as

$$\begin{aligned}
\int_{\Omega^0} (\Delta \hat{\mathbf{e}}_0)^T \hat{\mathbf{C}} \delta(\Delta \hat{\mathbf{e}}_0) d\Omega^0 + \int_{\Omega^0} \left(\hat{\mathbf{S}}_0^t \right)^T \delta(\Delta \hat{\boldsymbol{\eta}}_0) d\Omega^0 &= \int_{\Omega^0} \left(\hat{\mathbf{b}}_0^{t+\Delta t} \right)^T \delta \hat{\mathbf{u}} d\Omega^0 + \int_{\Gamma^0} \left(\hat{\mathbf{t}}_0^{t+\Delta t} \right)^T \delta \hat{\mathbf{u}} d\Gamma^0 \\
&+ \int_{\Omega^0} (\Delta \hat{\mathbf{e}}_0^V)^T \hat{\mathbf{C}} \delta(\Delta \hat{\mathbf{e}}_0) d\Omega^0 + \int_{\Omega^0} (\Delta \hat{\mathbf{e}}_0^{HT})^T \hat{\mathbf{C}} \delta(\Delta \hat{\mathbf{e}}_0) d\Omega^0 \\
&- \int_{\Omega^0} \left(\hat{\mathbf{S}}_0^t \right)^T \delta(\Delta \hat{\mathbf{e}}_0) d\Omega^0
\end{aligned} \tag{9.18}$$

where $\hat{\mathbf{C}}$ is the elastic constitutive matrix and the strain increment vectors are defined by

$$\Delta \hat{\mathbf{e}}_0 = \begin{bmatrix} \Delta e_{011} & \Delta e_{022} & \Delta e_{033} & 2\Delta e_{023} & 2\Delta e_{013} & 2\Delta e_{012} \end{bmatrix}^T \tag{9.19}$$

$$\Delta \hat{\boldsymbol{\eta}}_0 = \begin{bmatrix} \Delta \eta_{011} & \Delta \eta_{022} & \Delta \eta_{033} & 2\Delta \eta_{023} & 2\Delta \eta_{013} & 2\Delta \eta_{012} \end{bmatrix}^T$$

with $\Delta \hat{\mathbf{E}}_0 = \Delta \hat{\mathbf{e}}_0 + \Delta \hat{\boldsymbol{\eta}}_0$. Similar definitions are employed for the viscoelastic and hygrothermal strain increment vectors $\Delta \hat{\mathbf{e}}_0^v$ and $\Delta \hat{\mathbf{e}}_0^{HT}$. Using the interpolation functions to express the displacements and increment displacements in (9.10), we obtain the variations

$$\delta(\Delta \hat{\mathbf{e}}_0) = \mathbf{B}_L \delta \left(\Delta \tilde{\mathbf{u}}^{(e)} \right) \qquad \delta(\Delta \hat{\boldsymbol{\eta}}_0) = \mathbf{B}_{NL} \delta \left(\Delta \tilde{\mathbf{u}}^{(e)} \right) \tag{9.20}$$

where $\mathbf{B}_L$ and $\mathbf{B}_{NL}$ are the linear and nonlinear strain–displacement matrices [1, 7]. $\delta \left(\Delta \tilde{\mathbf{u}}^{(e)} \right)$ is the variation in the nodal displacement increment vector of the element.

Introducing the strain–displacement relations into (9.18) and using the interpolation functions to express the displacements appearing in this equation, we obtain the following incremental equilibrium relationship for an element (3)

$$\left(\mathbf{k}_L^t + \mathbf{k}_{NL}^t\right)\Delta\tilde{\mathbf{u}}^{(e)} = \mathbf{r}^{t+\Delta t} - \mathbf{f}_0^t + \Delta\mathbf{f}^V + \Delta\mathbf{f}^{HT} \tag{9.21}$$

being $\Delta\tilde{\mathbf{u}}^{(e)}$ and $\mathbf{r}^{t+\Delta t}$ the vector of nodal displacement increments and the vector of external nodal loading at time $t + \Delta t$ respectively, and

$$\mathbf{k}_L^t = \int_{\Omega^{0(e)}} \left(\mathbf{B}_L^t\right)^T \hat{\mathbf{C}} \mathbf{B}_L^t d\Omega^{0(e)} \quad \text{(linear stiffness matrix at time } t) \tag{9.22}$$

$$\mathbf{k}_{NL}^t = \int_{\Omega^{0(e)}} \left(\mathbf{B}_{NL}^t\right)^T \hat{\mathbf{S}}_0^t \mathbf{B}_{NL}^t d\Omega^{0(e)} \quad \text{(nonlinear stiffness matrix at time } t) \tag{9.23}$$

$$\mathbf{f}_0^t = \int_{\Omega^{0(e)}} \left(\mathbf{B}_L^t\right)^T \hat{\mathbf{S}}_0^t d\Omega^{0(e)} \tag{9.24}$$
(vector of nodal forces equivalent to the element stresses at time t)

$$\Delta\mathbf{f}^V = \int_{\Omega^{o(e)}} \left(\mathbf{B}_L^t\right)^T \hat{\mathbf{C}} \Delta\hat{\mathbf{e}}^V d\Omega^{o(e)} \quad \text{(viscoelastic load increment vector)} \tag{9.25}$$

$$\Delta\mathbf{f}^{HT} = \int_{\Omega^{o(e)}} \left(\mathbf{B}_L^t\right)^T \hat{\mathbf{C}} \Delta\hat{\mathbf{e}}^{HT} d\Omega^{o(e)} \quad \text{(hygrothermal load increment vector)} \tag{9.26}$$

In these last equations, the integrals are determined on the element domain in the initial configuration $\Omega^{0(e)}$. The matrices $\mathbf{B}_L^t$ and $\mathbf{B}_{NL}^t$ are the linear and nonlinear strain displacement matrices at time t, respectively. The present approach, for which the kinematic and static variables and integration domains are referred to the initial configuration, is known as total Lagrangian formulation. An alternative and equivalent approach consists of the updated Lagrangian formulation that, for each incremental step $t + \Delta t$, adopts $\mathbf{C}^t$ as reference configuration [1].

For the case of small displacements, the incremental equilibrium equation (9.21) becomes

$$\mathbf{k}_L^t \Delta\tilde{\mathbf{u}}^{(e)} = \mathbf{r}^{t+\Delta t} - \mathbf{r}^t + \Delta\mathbf{f}^V + \Delta\mathbf{f}^{HT} \tag{9.27}$$

9.4 Numerical Solution of the Equilibrium Equation

The numerical solution of the geometrically nonlinear problem (9.21) can be obtained using an iterative procedure in which the element equilibrium equation at time $t + \Delta t$ is given by

$$\left(\mathbf{k}_L^{t+\Delta t(i-1)} + \mathbf{k}_{NL}^{t+\Delta t(i-1)} \right) \Delta \tilde{\mathbf{u}}^{(e)(i)} = \mathbf{r}^{t+\Delta t(i)} - \mathbf{f}_0^{t+\Delta t(i-1)} + \Delta \mathbf{f}^{V(i)} + \Delta \mathbf{f}^{HT(i)} \quad (9.28)$$

where the superscripts i and $i-1$ indicate iterative steps. In this iterative approach, the element viscoelastic and hygrothermal load increment vectors, $\Delta \mathbf{f}^{V(i)}$ and $\Delta \mathbf{f}^{HT(i)}$, are taken as null for $i \geq 2$. For the first iteration $i = 1$, this last vector is computed by using (9.26), with

$$\Delta \hat{\mathbf{e}}^{HT(1)} = \alpha \Delta \Theta^{(1)} + \beta \Delta H^{(1)} \quad (9.29)$$

being α and β the vectors of the temperature expansion and hygroscopic expansion coefficients, respectively. $\Delta \Theta^{(1)}$ and $\Delta H^{(1)}$ are the temperature and moisture changes, respectively, for the first iterative step at time $t + \Delta t$. The element viscoelastic load increment vector $\Delta \mathbf{f}^{V(1)}$ is obtained for the first iteration at time $t + \Delta t$ using the viscoelastic strains computed by the equilibrated stresses corresponding to time t (see Chap. 3).

For an assemblage of finite elements, the global equilibrium equation can be written as

$$\left(\mathbf{K}_L^{t+\Delta t(i-1)} + \mathbf{K}_{NL}^{t+\Delta t(i-1)} \right) \Delta \tilde{U}^{(i)} = \mathbf{R}^{t+\Delta t(i)} - \mathbf{F}_0^{t+\Delta t(i-1)} + \Delta \mathbf{F}^{V(i)} + \Delta \mathbf{F}^{HT(i)} \quad (9.30)$$

where the variables have analogous meanings to those appearing in element equilibrium equation (9.28), but referred to the global coordinates. An alternative form of writing this global equilibrium equation is

$$\left(\mathbf{K}_L^{t+\Delta t(i-1)} + \mathbf{K}_{NL}^{t+\Delta t(i-1)} \right) \Delta \tilde{U}^{(i)} = \Delta \lambda^{(i)} \bar{\mathbf{P}} + \mathbf{F}_d^{t+\Delta t(i-1)} + \Delta \mathbf{F}^{V(i)} + \Delta \mathbf{F}^{HT(i)} \quad (9.31)$$

where $\Delta \lambda^{(i)}$ is the loading factor corresponding to the iteration i at time $t + \Delta t$, $\bar{\mathbf{P}}$ is the reference load vector and $\mathbf{F}_d^{t+\Delta t(i-1)}$ is the unbalanced force vector at the iteration $(i-1)$ of the step $t + \Delta t$. Using (9.30), the vector $\Delta \mathbf{F}^{HT(i)}$, for each time step, must be computed for the temperature and moisture increments $\Delta \lambda^{(1)} \bar{\Theta}$ and $\Delta \lambda^{(1)} \bar{H}$, being $\bar{\Theta}$ and $\bar{H}$ reference temperature and moisture values, respectively.

As solution algorithm to solve (9.31) we may use, for instance, the well-known Newton–Raphson method [2]. In this method, the loading factor value $\Delta \lambda^{(i)}$ is adopted in the beginning of the first iteration ($i = 1$) of each incremental step and is null for $i \geq 2$. One limitation of the Newton–Raphson method is the numerical instability that occurs near the limit points. To overcome this problem, we may use a displacement control algorithm, such as the Generalized Displacement Control Method [6]. The application of this method to viscoelastic problems can be found in Pavan et al. [5] and Oliveira and Creus [4].

9.5 Procedures of the Viscoelastic Finite Element Analysis

The implementation of the above geometrically nonlinear finite element formulation for the analysis of viscoelastic problems consists of the following main steps:

(1) Input the data for geometry, control parameters, mesh discretization, boundary conditions;

(2) Input the mechanical loads, temperature and moisture changes in each loading stage;

(3) Input the material properties corresponding to the temperature and moisture values;

(4) Assemble the strain–displacement matrices $(\mathbf{B}_L^{t(i)}, \mathbf{B}_{NL}^{t(i)})$ in the integration points of the elements;

(5) Assemble the element stiffness matrices $(\mathbf{k}_L^{t(i)}, \mathbf{k}_{NL}^{t(i)})$ and global stiffness matrices $(\mathbf{K}_L^{t(i)}, \mathbf{K}_{NL}^{t(i)})$;

(6) If there are temperature and moisture changes in the current loading stage and $i = 1$, assemble the element and global hygrothermal load increment vectors $(\Delta \mathbf{f}^{HT(1)}, \Delta \mathbf{F}^{HT(1)})$. For $i \geq 2$, $\Delta \mathbf{f}^{HT(i)} = 0$ and $\Delta \mathbf{F}^{HT(i)} = 0$;

(7) If the external loading was already applied at the current loading stage, assemble the element and global viscoelastic load increment vectors corresponding to the time interval of the incremental step $(\Delta \mathbf{f}^{V(1)}, \Delta \mathbf{F}^{V(1)})$. The viscoelastic strains can be computed by the state variables approach, as seen in Chaps. 3 and 4. For $i \geq 2$, $\Delta \mathbf{f}^{V(i)} = 0$ and $\Delta \mathbf{F}^{V(i)} = 0$;

(8) Compute the nodal displacement increments $\Delta \tilde{\mathbf{U}}^{(i+1)}$;

(9) Update the nodal displacement $\tilde{\mathbf{U}}^{t(i+1)} = \tilde{\mathbf{U}}^{t(i)} + \Delta \tilde{\mathbf{U}}^{(i+1)}$ and nodal coordinates;

(10) Assemble the strain–displacement matrices $(\mathbf{B}_L^{t(i+1)}, \mathbf{B}_{NL}^{t(i+1)})$ for the integration points of the elements in the updated configuration;

(11) Compute the stresses in the element integration points and vectors of nodal forces equivalent to these stresses (9.24), $\mathbf{F}_0^{t(i+1)}$, for the updated configuration;

(12) Determine the unbalanced force vector $\mathbf{F}_d^{t(i+1)}$;

(13) If the convergence criterion is not satisfied, then, do $i = i+1$ and return to step 4;

(14) If the convergence criterion is satisfied, two additional conditions must be checked: (a) if the time interval corresponding to the current loading stage is not complete, do $t + \Delta t$, $i = 1$ and go to step 4; (b) if the time interval is complete, then return to the new loading stage (step 2), if it exists, continuing the analysis.

Applications of these procedures to the analysis of viscoelastic laminated plates and shells may be found in Marques and Creus [3] and applications to viscoelastic thin-walled composite beams in Oliveira and Creus [4].

References

1. K.-J. Bathe, *Finite Element Procedures in Engineering Analysis* (Prentice-Hall, Inc Englewood Cliffs, New Jersey, 1996)
2. M.A. Crisfield, *Non-linear finite element analysis of solids and structures*, vol. 1 (John Wiley & Sons Ltd, West Sussex, 2003)
3. S.P.C. Marques, G.J. Creus, Geometrically nonlinear finite elements analysis of viscoelastic composite materials under mechanical and hygrothermal loads. Comput. Struct. **53**, 449–456 (1994)
4. B.F. Oliveira, G.J. Creus, Nonlinear viscoelastic analysis of thin-walled beams in composite material. Thin-Walled Struct. **41**, 957–971 (2003)
5. R.C. Pavan, B.F. Oliveira, S. Maghous, G.J. Creus, A model for anisotropic viscoelastic damage in composites. Compos. Struct. **92**, 1223–1228 (2010)
6. Y.B. Yang, M.S. Shieh, Solution method for nonlinear problems with multiple critical points. AIAA J. **28**(12), 2110–2116 (1990)
7. O.C. Zienkiewicz, R.L. Taylor, *RL Finite Element Method for Solid and Structural Mechanics* (Elsevier Butterworth-Heinemann, Jordan Hill, Oxford, 2005)

Chapter 10
The Boundary Element Method for Viscoelasticity Problems

The Boundary Element Method (BEM) is derived through the discretization of an integral equation (the classical Somigliana identity, first published in 1886). An interesting account of BEM early development may be found in [2]. This formulation can only be derived for certain classes of problems and hence, is not as widely applicable as the finite element method. However, when applicable, it often results in numerical methods that are easier to use and computationally more efficient. The advantages of the BEM arise from the fact that only the boundary of the domain requires sub-division. In cases where the domain is exterior to the boundary (e.g. the atmosphere surrounding an airplane, the soil surrounding a tunnel, the material surrounding a crack tip) the advantages of the BEM are even greater as the equation governing the infinite domain is reduced to an equation over the (finite) boundary. In this chapter we shortly review two alternative procedures for the solution of problems in linear viscoelasticity: the solution in the Laplace transformed domain and the use of a general inelastic formulation. For the latter, we make reference to the use of the Dual Reciprocity Method (DRM) that allows a pure boundary formulation.

10.1 Linear Elastic Problems and Somigliana Identity

We begin with a short summary of the classical boundary element formulation [1]. The boundary element method for linear elasticity may be established beginning with the Somigliana identity. Let us consider a body of volume Ω and surface Γ subjected to body forces b_k and surface forces p_i (following a tradition in the area, p in place of t will be used in this Chapter to denote tractions). Then, the Somigliana identity

S. P. C. Marques and G. J. Creus, *Computational Viscoelasticity*,
SpringerBriefs in Computational Mechanics, DOI: 10.1007/978-3-642-25311-9_10,
© The Author(s) 2012

$$u_l^i + \int_\Gamma p_{lk}^* u_k d\Gamma = \int_\Gamma u_{lk}^* p_k d\Gamma + \int_\Omega u_{lk}^* b_k d\Omega \quad (10.1)$$

gives the value of the displacements at any internal points in terms of the boundary values of u_k and p_k, the domain forces b_k and the fundamental solutions u_{lk}^* and p_{lk}^*. p_{lk}^* are the tractions in the k direction due to a unit force at i acting in the l direction, and u_{lk}^* are the displacements in the k direction due to a unit force at i on the l direction. An updated derivation of the Somigliana identity may be found in [1], where (10.1) is obtained by reciprocity with a singular solution of the Navier equation for body force components modeled as unit point loads

$$G u_{l,kk}^* + \frac{G}{1 - 2v} u_{k,kl}^* + \Delta^i e_l = 0 \quad (10.2)$$

where Δ^i represents the Dirac delta function at i. For a boundary point, (10.1) transforms to

$$c_{lk}^i u_l^i + \int_\Gamma p_{lk}^* u_k d\Gamma = \int_\Gamma u_{lk}^* p_k d\Gamma + \int_\Omega u_{lk}^* b_k d\Omega \quad (10.3)$$

where the integrals are in the sense of Cauchy principal value. For Γ smooth at point i it is $c_{lk}^i = \delta_{lk}/2$.

10.1.1 Boundary Element Formulation for the Linear Elastic Case

In order to obtain a numerical procedure, the boundary is discretized in elements, over which displacements and tractions are expressed in terms of their values at the nodal points. Using now matrix notation,

$$\mathbf{u} = \mathbf{\Phi} \mathbf{u}^j \quad \mathbf{p} = \mathbf{\Phi} \mathbf{p}^j \quad (10.4)$$

where $\mathbf{u}^j$ and $\mathbf{p}^j$ are the element nodal displacements and tractions and the interpolation functions $\mathbf{\Phi}$ are the standard finite element type functions. Then, writing (10.3) in matrix form we have

$$\mathbf{c}^i \mathbf{u}^i + \int_\Gamma \mathbf{p}^* \mathbf{u} d\Gamma = \int_\Gamma \mathbf{u}^* \mathbf{p} d\Gamma + \int_\Omega \mathbf{u}^* \mathbf{b} d\Omega \quad (10.5)$$

and using (10.4)

$$\mathbf{c}^i \mathbf{u}^i + \sum_{j=1}^N \left\{ \int_{\Gamma_j} \mathbf{p}^* \mathbf{\Phi} d\Gamma \right\} \mathbf{u}^j = \sum_{j=1}^N \left\{ \int_{\Gamma_j} \mathbf{u}^* \mathbf{\Phi} d\Gamma \right\} \mathbf{p}^j + \sum_{s=1}^M \left\{ \int_{\Omega_s} \mathbf{u}^* \mathbf{b} d\Omega \right\} \quad (10.6)$$

The sum from $j = 1$ to N indicate summation over all the N elements, Γ_j is the surface of element j and $\mathbf{u}^j$ and $\mathbf{p}^j$ the corresponding displacements and tractions. The domain was divided into M internal cells of volume Ω_s over which the body forces integral have to be computed. After integration we have for a given node i

$$\mathbf{c}^i\mathbf{u}^i + \sum_{j=1}^{NE} \mathbf{H}^{ij}\mathbf{u}^j = \sum_{j=1}^{NE} \mathbf{G}^{ij}\mathbf{p}^j + \sum_{s=1}^{M} \mathbf{B}^{is} \tag{10.7}$$

The contribution for all the NE nodes may be written in matrix form

$$\mathbf{HU} = \mathbf{GP} + \mathbf{B} \tag{10.8}$$

After the boundary conditions are introduced, all unknowns are set into a vector $\mathbf{X}$ leading to a system of equations

$$\mathbf{AX} = \mathbf{F} \tag{10.9}$$

10.2 Viscoelastic Solutions in the Laplace Transform Domain

If the correspondence principle (see Chap. 5) is applied to the quasi-static problem, the relevant boundary integral equation in the Laplace transformed domain is written

$$c_{lk}^i \bar{u}_l^i(s) + \int_\Gamma p_{lk}^*(s)\bar{u}_k(s)d\Gamma = \int_\Gamma u_{lk}^*(s)\bar{p}_k(s)d\Gamma + \int_\Omega u_{lk}^*(s)\bar{b}_k(s)d\Omega \tag{10.10}$$

where now $u_{lk}^*(s)$ and $p_{lk}^*(s)$ are the elastic fundamental solutions for displacements and tractions in which the elastic constants have been replaced by the corresponding functions in the transformed space according to Sect. 5.2. A discussion of this type of approach may be found in Syngellakis [9], Gaul and Schanz [3]. The main difficulty is the inversion from the Laplace to the real (time) domain.

10.3 Formulation Considering Inelastic Strains

The general boundary integral equation including the effect of inelastic strains may be written in incremental form as [1]

$$c_{lk}^i \dot{u}_l^i = \int_\Gamma u_{lk}^*(\dot{p}_k + \dot{p}_k^v)d\Gamma - \int_\Gamma p_{lk}^* u_k d\Gamma + \int_\Omega u_{lk}^*(\dot{b}_k + \dot{b}_k^v)d\Omega \tag{10.11}$$

where

$$\dot{p}^v_i = \dot{\sigma}^v_{ij} n_j; \quad \dot{b}^v_i = -\dot{\sigma}^v_{ij,j}; \quad \dot{\sigma}^v_{ij} = E_{ijkl} \dot{\varepsilon}^v_{kl} \tag{10.12}$$

and ε^v_{ij} is the deferred part of strain as defined in Chap. 2. Equation (10.11) is known as the *pseudo-surface traction, pseudo-body force approach*; the inelastic forces are included adding $\dot{p}^v_k$ to $\dot{p}_k$ in the surface traction boundary integral and $\dot{b}^v_k$ to $\dot{b}_k$ in the body force domain integral. This formulation has been applied to time-dependent problems by a series of authors; see for example Brebbia et al. [1]. The domain integral has to be computed using cells defined over the domain. There are alternatives that avoid the domain integration, one of which is the Dual Reciprocity Formulation (DRM) [6].

10.3.1 DRM Applied to Viscoelasticity

With reference to (10.11), we define $\dot{w}^v$ so that $\dot{w}^v_{,i} = \dot{b}^v_j$. Using the DRM strategy, we expand $\dot{w}^v$ as the sum of known approximating functions with initially unknown coefficients

$$\dot{w}^v \simeq \sum_{j=1}^{M} f^m \dot{\alpha}^m \tag{10.13}$$

where M is the number of DRM collocation points. Differentiating (10.13) we obtain

$$\dot{w}^v_j = \dot{b}^v_j \simeq \sum_{j=1}^{M} f^m_j \dot{\alpha}^m \tag{10.14}$$

Considering (10.11) and making the regular body forces $b_j = 0$, we can now substitute $\dot{b}^v_j$ given by (10.14) obtaining

$$c^i_{lk} \dot{u}^i_l = \int_\Gamma u^*_{lk} (\dot{p}_k + \dot{p}^v_k) d\Gamma - \int_\Gamma p^*_{lk} \dot{u}_k d\Gamma + \sum_{m=1}^{M} \left(\int_\Omega u^*_{lk} f^m_{,k} \right) \dot{\alpha}^m \tag{10.15}$$

The DRM particular solutions $\hat{u}_j$ should satisfy the Navier equation

$$G\hat{u}^j_{k,ll} + \frac{G}{1-2v} \hat{u}^j_{l,lk} = f^j_{,k} \tag{10.16}$$

Taking the domain term to the boundary with DRM we obtain

$$c_{lk}^i \dot{u}_l^i = \int_\Gamma u_{lk}^*(\dot{p}_k + \dot{p}_k^v)d\Gamma - \int_\Gamma p_{lk}^* \dot{u}_k d\Gamma + \sum_{m=1}^{M} \left(c_{lk}\hat{u}_l^m - \int_\Gamma u_{lk}^* \hat{p}_l^m d\Gamma + \int_\Gamma p_{lk}^* \hat{u}_l^m d\Gamma \right) \dot{\alpha}^m$$

(10.17)

After discretization and approximation of the above equation to all boundary nodes, the following system of equations is obtained

$$\mathbf{H\dot{U}} - \mathbf{G\dot{P}} = (\mathbf{H\hat{U}} - \mathbf{G\hat{P}})\dot{\boldsymbol{\alpha}}$$

(10.18)

or, substituting from (10.13) $\dot{\boldsymbol{\alpha}} = \boldsymbol{F}^{-1}\dot{\mathbf{w}}$

$$\mathbf{H\dot{U}} - \mathbf{G\dot{P}} = (\mathbf{H\hat{U}} - \mathbf{G\hat{P}})\boldsymbol{F}^{-1}\dot{\mathbf{w}}^v$$

(10.19)

or

$$\mathbf{H\dot{U}} - \mathbf{G\dot{P}} = \dot{\mathbf{D}}$$

(10.20)

Applying the usual BEM procedure we set the system of equations in the form

$$\mathbf{A\dot{X}} = \dot{\mathbf{Y}} + \dot{\mathbf{D}}$$

(10.21)

From its solution we obtain $\dot{u}_i, \dot{p}_i$, and we can determine the boundary and internal stress tensors and advance in time. Additional and numerical examples may be found in [8].

10.4 Other Procedures

Other different and complementary procedures may be seen in Liu and Antes [4]; Mesquita and Coda [5]; Schanz and Antes [7].

References

1. C.A. Brebbia, J.C.F. Telles, L.C. Wrobel, *Boundary Element Technique* (Springer, Berlin, 1984)
2. A.H.D. Cheng, D.T. Cheng, Heritage and early history of the boundary element method. Eng. Anal. Boundary Elem **29**, 268–302 (2005)
3. L. Gaul, M. Schanz, A comparative study of three boundary element approaches to calculate the transient response of viscoelastic solids with unbounded domains. Comput. Methods Appl. Mech. Engrg. **179**, 111–123 (1999)
4. Y. Liu, H. Antes, Application of visco-elastic boundary element method to creep problems in chemical engineering structures. Int. J. Press. Vessel. Pip. **70**(1), 27–31 (1997)
5. A.D. Mesquita, H.B. Coda, Boundary integral equation method for general viscoelastic analysis. Int. J. Solids Struct. **39**, 2643–2664 (2002)

6. P.W. Partridge, C.A. Brebbia, L.C. Wrobel, *The Dual Reciprocity Boundary Element Method* (Computational Mechanics Publication, Southampton, 1992)
7. M. Schanz, H. Antes, A new visco- and elastodynamic time domain boundary element formulation. Comput. Mech. **20**(5), 452–459 (1997)
8. B. Sensale, P.W. Partridge, G.J. Creus, General boundary elements solution for ageing viscoelastic structures. Int. J. Numer. Meth. Engng. **50**, 1455–1468 (2001)
9. S. Syngellakis, Boundary element methods for polymer analysis. Eng. Anal. Boundary Elem. **27**, 125–135 (2003)

Chapter 11
Viscoelastic Finite Volume Formulation

The finite-volume theory is a well-known technique frequently used for analysis of boundary-value problems in fluid mechanics [7]. Its excellent performance has motivated attempts to extend it to solid mechanics problems. Thus, in the past two decades, several authors presented formulations based on this technique. Here, we present one of these finite-volume formulations, known as the *Parametric Finite-Volume Formulation*. It uses the Finite Volume Direct Averaged Method—FVDAM [1] as a basis and is summarized for the case of linear elastic problems in Cavalcante et al. [2, 3]. An extension of the *Parametric Finite-Volume Formulation* in order to include linear viscoelastic effect, here presented, can be found in a more detailed form in Escarpini Filho [5].

11.1 Parametric Finite Volume Formulation: Background

In the two dimensional parametric finite-volume formulation the domain occupied by the solid is discretized into quadrilateral subvolumes. The formulation is based on the mapping of a reference square subvolume onto each one of these quadrilateral subvolume resident on the actual solid domain, as shown in Fig. 11.1 [2]. The mapping of the point (η, ξ) in the reference subvolume to the corresponding actual point (x, y) is defined by

$$
\begin{aligned}
x(\eta, \xi) &= N_1(\eta, \xi)x_1 + N_2(\eta, \xi)x_2 + N_3(\eta, \xi)x_3 + N_4(\eta, \xi)x_4 \\
y(\eta, \xi) &= N_1(\eta, \xi)y_1 + N_2(\eta, \xi)y_2 + N_3(\eta, \xi)y_3 + N_4(\eta, \xi)y_4
\end{aligned}
\tag{11.1}
$$

where N_i are the shape functions given by $N_1 = (1 - \eta)(1 - \xi)/4$, $N_2 = (1 + \eta)(1 - \xi)/4$, $N_3 = (1 + \eta)(1 + \xi)/4$ and $N_4 = (1 - \eta)(1 + \xi)/4$. x_i and y_i are the Cartesian coordinates of the i-*th* node of the quadrilateral subvolume. Here, we use x and y as geometrical coordinates instead of x_1 and x_2.

S. P. C. Marques and G. J. Creus, *Computational Viscoelasticity*,
SpringerBriefs in Computational Mechanics, DOI: 10.1007/978-3-642-25311-9_11,
© The Author(s) 2012

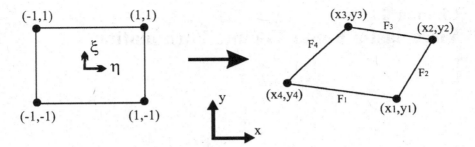

Fig. 11.1 Mapping of the reference square subvolume in the η-ξ plane onto a quadrilateral subvolume in the x–y plane of the actual microstructure

The displacement components in the reference subvolume are approximated by second order polynomials as follows

$$u_i = U_{i(00)} + \eta U_{i(10)} + \xi U_{i(01)} + \frac{1}{2}\left(3\eta^2 - 1\right)U_{i(20)} + \frac{1}{2}\left(3\xi^2 - 1\right)U_{i(02)} \quad (11.2)$$

where $i = 1, 2$ and $U_{i()}$ are the unknown displacement field coefficients. Thus, for two dimensional problems, we have ten unknown coefficients to be determined.

The surface-averaged partial derivative of the displacement components on the subvolume faces $F_k(k = 1, 2, 3, 4)$ are defined by

$$\left\langle \frac{\partial u_i}{\partial \eta} \right\rangle\bigg|_{k=1,3} = \frac{1}{2}\int_{-1}^{+1}\frac{\partial u_i}{\partial \eta}d\eta \quad \left\langle \frac{\partial u_i}{\partial \xi} \right\rangle\bigg|_{k=1,3} = \frac{1}{2}\int_{-1}^{+1}\frac{\partial u_i}{\partial \xi}d\eta \quad (11.3)$$

$$\left\langle \frac{\partial u_i}{\partial \eta} \right\rangle\bigg|_{k=2,4} = \frac{1}{2}\int_{-1}^{+1}\frac{\partial u_i}{\partial \eta}d\xi \quad \left\langle \frac{\partial u_i}{\partial \xi} \right\rangle\bigg|_{k=2,4} = \frac{1}{2}\int_{-1}^{+1}\frac{\partial u_i}{\partial \xi}d\xi$$

which, using (11.2), can be expressed as functions of the displacement coefficients.

For the face F_k, the partial derivatives with respect to the Cartesian coordinates are related to the partial derivatives given in (11.3) by

$$\begin{bmatrix} \left\langle \frac{\partial u_1}{\partial x} \right\rangle \\ \left\langle \frac{\partial u_1}{\partial y} \right\rangle \\ \left\langle \frac{\partial u_2}{\partial x} \right\rangle \\ \left\langle \frac{\partial u_2}{\partial y} \right\rangle \end{bmatrix}^{(k)} = \begin{bmatrix} \hat{\mathbf{J}} & \mathbf{0} \\ \mathbf{0} & \hat{\mathbf{J}} \end{bmatrix} \begin{bmatrix} \left\langle \frac{\partial u_1}{\partial \eta} \right\rangle \\ \left\langle \frac{\partial u_1}{\partial \xi} \right\rangle \\ \left\langle \frac{\partial u_2}{\partial \eta} \right\rangle \\ \left\langle \frac{\partial u_2}{\partial \xi} \right\rangle \end{bmatrix}^{(k)} \quad (11.4)$$

being $\hat{\mathbf{J}}$ the inverse of the volume-averaged Jacobian matrix $\bar{\mathbf{J}}$ defined by

$$\bar{\mathbf{J}} = \frac{1}{4} \int_{-1}^{+1} \int_{-1}^{+1} \mathbf{J} d\eta d\xi \tag{11.5}$$

and

$$\mathbf{J} = \begin{bmatrix} \frac{\partial x}{\partial \eta} & \frac{\partial y}{\partial \eta} \\ \frac{\partial x}{\partial \xi} & \frac{\partial y}{\partial \xi} \end{bmatrix} \tag{11.6}$$

Using the strain–displacement relation (5.39) with $x_1 = x$ and $x_2 = y$, we can write the surface-averaged strain vector for the face F_k as

$$\begin{bmatrix} \langle \varepsilon_{11} \rangle \\ \langle \varepsilon_{22} \rangle \\ \langle \varepsilon_{12} \rangle \end{bmatrix}^{(k)} = \bar{\mathbf{L}} \begin{bmatrix} \left\langle \frac{\partial u_1}{\partial x} \right\rangle \\ \left\langle \frac{\partial u_1}{\partial y} \right\rangle \\ \left\langle \frac{\partial u_2}{\partial x} \right\rangle \\ \left\langle \frac{\partial u_2}{\partial y} \right\rangle \end{bmatrix}^{(k)} \tag{11.7}$$

being

$$\bar{\mathbf{L}} = \begin{bmatrix} 1 & 0 & 0 & 0 \\ 0 & 0 & 0 & 1 \\ 0 & 1/2 & 1/2 & 0 \end{bmatrix} \tag{11.8}$$

For elastic analyses, the surface-averaged stress vector on the subvolume face F_k is given by

$$\begin{bmatrix} \langle \sigma_{11} \rangle \\ \langle \sigma_{22} \rangle \\ \langle \sigma_{12} \rangle \end{bmatrix}^{(k)} = \hat{\mathbf{C}} \begin{bmatrix} \langle \varepsilon_{11} \rangle \\ \langle \varepsilon_{22} \rangle \\ \langle \varepsilon_{12} \rangle \end{bmatrix}^{(k)} \tag{11.9}$$

being $\hat{\mathbf{C}}$ the material elastic constitutive matrix. Applying the Cauchy relation, we have the surface-averaged tractions as follows

$$\begin{bmatrix} t_1^{(k)} \\ t_2^{(k)} \end{bmatrix} = \begin{bmatrix} n_1^{(k)} & 0 & n_2^{(k)} \\ 0 & n_2^{(k)} & n_1^{(k)} \end{bmatrix} \begin{bmatrix} \langle \sigma_{11} \rangle \\ \langle \sigma_{22} \rangle \\ \langle \sigma_{12} \rangle \end{bmatrix}^{(k)} \tag{11.10}$$

where $n_i^{(k)}$ stand for the components of the outward unit normal vector to the face F_k. Substituting (11.9) into (11.10) and using (11.7) together with (11.2)–(11.4), we derive the relation between the surface-averaged tractions on the subvolume faces and unknown displacement coefficients

$$\mathbf{t} = \bar{\mathbf{A}} \mathbf{U} \tag{11.11}$$

where

$$\mathbf{t} = \left[\, t_1^{(1)} \quad t_2^{(1)} \quad t_1^{(2)} \quad t_2^{(2)} \quad t_1^{(3)} \quad t_2^{(3)} \quad t_1^{(4)} \quad t_2^{(4)} \,\right]^T \tag{11.12}$$

$$\mathbf{U} = \left[\, U_{1(10)} \quad U_{1(01)} \quad U_{1(20)} \quad U_{1(02)} \quad U_{2(10)} \quad U_{2(01)} \quad U_{2(20)} \quad U_{2(02)} \,\right]^T \tag{11.13}$$

and $\bar{\mathbf{A}} = \mathbf{DHLBA}$. These matrices are found in Appendix C.

Considering body forces b_1 and b_2, the differential equilibrium equations are given by

$$\frac{\partial \sigma_{11}}{\partial x} + \frac{\partial \sigma_{21}}{\partial y} + b_1 = 0 \tag{11.14}$$

$$\frac{\partial \sigma_{12}}{\partial x} + \frac{\partial \sigma_{22}}{\partial y} + b_2 = 0$$

from which, after writing the stress derivatives in function of the displacement coefficients, the following relations can be obtained

$$\begin{aligned}
\left(\hat{C}_{11}\bar{J}_{11}^2 + \hat{C}_{33}\bar{J}_{21}^2\right)U_{1(20)} &+ \left(\hat{C}_{11}\bar{J}_{12}^2 + \hat{C}_{33}\bar{J}_{22}^2\right)U_{1(02)} \\
&+ \left(\hat{C}_{12} + \hat{C}_{33}\right)\left(\bar{J}_{11}\bar{J}_{21}U_{2(20)} + \bar{J}_{12}\bar{J}_{22}U_{2(02)}\right) \\
&= -\frac{b_1}{3}
\end{aligned} \tag{11.15}$$

$$\begin{aligned}
\left(\hat{C}_{21} + \hat{C}_{33}\right)\left(\bar{J}_{11}\bar{J}_{21}U_{1(20)} + \bar{J}_{12}\bar{J}_{22}U_{1(02)}\right) \\
&+ \left(\hat{C}_{22}\bar{J}_{21}^2 + \hat{C}_{33}\bar{J}_{11}^2\right)U_{2(20)} + \left(\hat{C}_{22}\bar{J}_{22}^2 + \hat{C}_{33}\bar{J}_{12}^2\right)U_{2(02)} \\
&= -\frac{b_2}{3}
\end{aligned}$$

Introducing the displacement approximations (11.2) into the definitions of the surface-averaged displacement components in each subvolume face $F_k(k = 1, 2, 3, 4)$

$$\bar{u}_i^{(k=1,3)} = \langle u_i \rangle|_{k=1,3} = \frac{1}{2}\int_{-1}^{+1} u_i d\eta \qquad \bar{u}_i^{(k=2,4)} = \langle u_i \rangle|_{k=2,4} = \frac{1}{2}\int_{-1}^{+1} u_i d\xi \tag{11.16}$$

we derive the following relation between these displacement components and the displacement coefficients

$$\mathbf{U} = \frac{1}{2}\begin{bmatrix} 0 & 1 & 0 & -1 & 0 & 0 & 0 & 0 \\ -1 & 0 & 1 & 0 & 0 & 0 & 0 & 0 \\ 0 & 1 & 0 & 1 & 0 & 0 & 0 & 0 \\ 1 & 0 & 1 & 0 & 0 & 0 & 0 & 0 \\ 0 & 0 & 0 & 0 & 0 & 1 & 0 & -1 \\ 0 & 0 & 0 & 0 & -1 & 0 & 1 & 0 \\ 0 & 0 & 0 & 0 & 0 & 1 & 0 & 1 \\ 0 & 0 & 0 & 0 & 1 & 0 & 1 & 0 \end{bmatrix}\begin{bmatrix} \bar{u}_1^{(1)} - U_{1(00)} \\ \bar{u}_1^{(2)} - U_{1(00)} \\ \bar{u}_1^{(3)} - U_{1(00)} \\ \bar{u}_1^{(4)} - U_{1(00)} \\ \bar{u}_2^{(1)} - U_{2(00)} \\ \bar{u}_2^{(2)} - U_{2(00)} \\ \bar{u}_2^{(3)} - U_{2(00)} \\ \bar{u}_2^{(4)} - U_{2(00)} \end{bmatrix},\tag{11.17}$$

being $\mathbf{U}$ the vector appearing in (11.11).
From (11.15) and (11.17), we can show that

$$\mathbf{U} = \bar{\mathbf{B}}\bar{\mathbf{u}} - \mathbf{N}\Phi^{-1}\Omega \tag{11.18}$$

where $\bar{\mathbf{u}} = \begin{bmatrix} \bar{u}_1^{(1)} & \bar{u}_2^{(1)} & \bar{u}_1^{(2)} & \bar{u}_2^{(2)} & \bar{u}_1^{(3)} & \bar{u}_2^{(3)} & \bar{u}_1^{(4)} & \bar{u}_2^{(4)} \end{bmatrix}^T$ is the subvolume surface averaged displacement vector,

$$\bar{\mathbf{B}} = \mathbf{P} - \mathbf{N}\Phi^{-1}\Theta\mathbf{M} \tag{11.19}$$

and

$$\Phi = \begin{bmatrix} \hat{C}_{11}\left(\bar{J}_{11}^2 + \bar{J}_{12}^2\right) + \hat{C}_{33}\left(\bar{J}_{21}^2 + \bar{J}_{22}^2\right) & \left(\hat{C}_{12} + \hat{C}_{33}\right)\left(\bar{J}_{11}\bar{J}_{21} + \bar{J}_{12}\bar{J}_{22}\right) \\ \left(\hat{C}_{12} + \hat{C}_{33}\right)\left(\bar{J}_{11}\bar{J}_{21} + \bar{J}_{12}\bar{J}_{22}\right) & \hat{C}_{22}\left(\bar{J}_{21}^2 + \bar{J}_{22}^2\right) + \hat{C}_{33}\left(\bar{J}_{11}^2 + \bar{J}_{12}^2\right) \end{bmatrix} \tag{11.20}$$

$$\Omega = \frac{1}{3}\begin{bmatrix} b_1 \\ b_2 \end{bmatrix} \tag{11.21}$$

$$\Theta^T = \frac{1}{2}\begin{bmatrix} \hat{C}_{11}\bar{J}_{11}^2 + \hat{C}_{33}\bar{J}_{21}^2 & \left(\hat{C}_{12} + \hat{C}_{33}\right)\bar{J}_{11}\bar{J}_{21} \\ \hat{C}_{11}\bar{J}_{12}^2 + \hat{C}_{33}\bar{J}_{22}^2 & \left(\hat{C}_{12} + \hat{C}_{33}\right)\bar{J}_{12}\bar{J}_{22} \\ \left(\hat{C}_{12} + \hat{C}_{33}\right)\bar{J}_{11}\bar{J}_{21} & \hat{C}_{22}\bar{J}_{21}^2 + \hat{C}_{33}\bar{J}_{11}^2 \\ \left(\hat{C}_{12} + \hat{C}_{33}\right)\bar{J}_{12}\bar{J}_{22} & \hat{C}_{22}\bar{J}_{22}^2 + \hat{C}_{33}\bar{J}_{12}^2 \end{bmatrix} \tag{11.22}$$

The matrices $\mathbf{P}$, $\mathbf{M}$ and $\mathbf{N}$ are presented in Appendix C.

Substituting (11.18) into (11.11) we obtain the relation between the surface-averaged displacements and the surface-averaged tractions

$$\mathbf{t} = \mathbf{k}\bar{\mathbf{u}} - \mathbf{t}_0 \tag{11.23}$$

where $\mathbf{k} = \bar{\mathbf{A}}\bar{\mathbf{B}}$ is the local stiffness matrix and $\mathbf{t}_0 = \bar{\mathbf{A}}\mathbf{N}\Phi^{-1}\Omega$ represents a traction vector related to the body forces.

The global stiffness matrix is generated using a procedure similar to that employed in finite element algorithms. However, in the *Parametric Finite Volume Formulation* the global system of equations is obtained applying both surface-averaged interfacial traction and displacement continuity conditions, followed by the prescribed boundary conditions. Each face of a quadrilateral subvolume has two degrees of freedom associated to two surface-averaged displacement components and two degrees of freedom related to two surface-averaged traction components. From this assemblage procedure, a global system of equations is obtained with the form

$$\mathbf{k}_G \bar{\mathbf{u}}_G - \mathbf{t}_{G0} = \mathbf{t}_G \qquad (11.24)$$

where $\mathbf{k}_G$ is the global stiffness matrix and the vector $\bar{\mathbf{u}}_G$ contains all the unknown interfacial and boundary surface-averaged displacements. $\mathbf{t}_G$ contains information on the surface-averaged tractions along the interfaces and the discretized boundary. $\mathbf{t}_{G0}$ is constituted by net contributions of the local vectors $\mathbf{t}_0$ along those interfaces and the discretized boundary.

Solving the system of equations (11.24), we obtain the surface-averaged displacement vector $\bar{\mathbf{u}}_G$ and, in sequel, the surface-averaged displacement vector $\bar{\mathbf{u}}$ of each subvolume. Afterwards, using (11.18) and (11.17), we determine all subvolume displacement coefficients and, subsequently, the subvolume displacement fields u_i given by (11.2). Then, utilizing the strain–displacement equations and the material constitutive relations, we obtain the subvolume strain and stress components.

It is worth mentioning that the imposition of the continuity of surface-averaged displacements and tractions used in the *Parametric Finite Volume Formulation* gives to this tool a particular efficiency for the analysis of structures constituted by heterogeneous materials [2–4].

11.2 Viscoelastic Parametric Finite Volume Formulation

For a viscoelastic subvolume, the incremental displacement field, corresponding to time interval $[t, t + \Delta t]$, is expressed by the following second order approximations

$$\Delta u_i = \Delta U_{i(00)} + \eta \Delta U_{i(10)} + \xi \Delta U_{i(01)} + \frac{1}{2}\left(3\eta^2 - 1\right)\Delta U_{i(20)} + \frac{1}{2}\left(3\xi^2 - 1\right)\Delta U_{i(02)}$$

$$(11.25)$$

where $\Delta U_{i()}$ are unknown coefficients. The increments of stress are related to increments of strain through the expression

$$\begin{bmatrix} \Delta\sigma_{11} \\ \Delta\sigma_{22} \\ \Delta\sigma_{12} \end{bmatrix} = \hat{\mathbf{C}} \left(\begin{bmatrix} \Delta\varepsilon_{11} \\ \Delta\varepsilon_{22} \\ \Delta\varepsilon_{12} \end{bmatrix} - \begin{bmatrix} \Delta\varepsilon_{11}^{HT} \\ \Delta\varepsilon_{22}^{HT} \\ \Delta\varepsilon_{12}^{HT} \end{bmatrix} - \begin{bmatrix} \Delta\varepsilon_{11}^{V} \\ \Delta\varepsilon_{22}^{V} \\ \Delta\varepsilon_{12}^{V} \end{bmatrix} \right) \qquad (11.26)$$

being $\Delta\varepsilon_{ij}^{HT}$ and $\Delta\varepsilon_{ij}^{V}$, the hygrothermal and viscoelastic strain increments, respectively. The surface-averaged incremental stress–strain relation for the subvolume face k can be expressed in the form

$$
\begin{bmatrix} \langle\Delta\sigma_{11}\rangle \\ \langle\Delta\sigma_{22}\rangle \\ \langle\Delta\sigma_{12}\rangle \end{bmatrix}^{(k)} = \hat{\mathbf{C}} \left(\begin{bmatrix} \langle\Delta\varepsilon_{11}\rangle \\ \langle\Delta\varepsilon_{22}\rangle \\ \langle\Delta\varepsilon_{12}\rangle \end{bmatrix} - \begin{bmatrix} \langle\Delta\varepsilon_{11}^{HT}\rangle \\ \langle\Delta\varepsilon_{22}^{HT}\rangle \\ \langle\Delta\varepsilon_{12}^{HT}\rangle \end{bmatrix} - \begin{bmatrix} \langle\Delta\varepsilon_{11}^{V}\rangle \\ \langle\Delta\varepsilon_{22}^{V}\rangle \\ \langle\Delta\varepsilon_{12}^{V}\rangle \end{bmatrix} \right)^{(k)}
\tag{11.27}
$$

Applying to this equation the same procedure used to derive (11.11), i.e., introducing Cauchy formula in (11.27) and using the relations between displacement increments and strain increments, we obtain the surface-averaged traction increment vector of a subvolume

$$
\Delta\mathbf{t} = \bar{\mathbf{A}}\Delta\mathbf{U} - \mathbf{DH}\big(\Delta\bar{\boldsymbol{\varepsilon}}^{HT} + \Delta\bar{\boldsymbol{\varepsilon}}^{V}\big)
\tag{11.28}
$$

where $\Delta\bar{\boldsymbol{\varepsilon}}^{HT}$ and $\Delta\bar{\boldsymbol{\varepsilon}}^{V}$ are the hygrothermal and viscoelastic surface-averaged strain increment vectors, respectively,

$$
\Delta\mathbf{t} = \left[\Delta t_1^{(1)}\ \Delta t_2^{(1)}\ \Delta t_1^{(2)}\ \Delta t_2^{(2)}\ \Delta t_1^{(3)}\ \Delta t_2^{(3)}\ \Delta t_1^{(4)}\ \Delta t_2^{(4)} \right]^{T}
\tag{11.29}
$$

$$
\Delta\mathbf{U} = \left[\Delta U_{1(10)}\ \Delta U_{1(01)}\ \Delta U_{1(20)}\ \Delta U_{1(02)}\ \Delta U_{2(10)}\ \Delta U_{2(01)}\ \Delta U_{2(20)}\ \Delta U_{2(02)} \right]^{T}
\tag{11.30}
$$

Similarly to (11.18), we may show that the incremental displacement coefficient vector $\Delta\mathbf{U}$ is related to the surface-averaged incremental displacement vector $\Delta\bar{\mathbf{u}}$ by

$$
\Delta\mathbf{U} = \bar{\mathbf{B}}\Delta\bar{\mathbf{u}} - \mathbf{N}\Phi^{-1}\Delta\boldsymbol{\Omega}
\tag{11.31}
$$

being the incremental body force vector

$$
\Delta\boldsymbol{\Omega} = \frac{1}{3}\begin{bmatrix} \Delta b_1 \\ \Delta b_2 \end{bmatrix}
\tag{11.32}
$$

and

$$
\Delta\bar{\mathbf{u}} = \left[\Delta\bar{u}_1^{(1)}\ \Delta\bar{u}_2^{(1)}\ \Delta\bar{u}_1^{(2)}\ \Delta\bar{u}_2^{(2)}\ \Delta\bar{u}_1^{(3)}\ \Delta\bar{u}_2^{(3)}\ \Delta\bar{u}_1^{(4)}\ \Delta\bar{u}_2^{(4)} \right]^{T}
\tag{11.33}
$$

Substituting (11.31) into (11.28), we obtain the local incremental relation

$$
\Delta\mathbf{t} = \mathbf{k}\Delta\bar{\mathbf{u}} - \Delta\mathbf{t}_0
\tag{11.34}
$$

where $\mathbf{k}$ is the same stiffness matrix appearing in (11.23) and

$$
\Delta\mathbf{t}_0 = \bar{\mathbf{A}}\mathbf{N}\Phi^{-1}\Delta\boldsymbol{\Omega} + \mathbf{DH}(\Delta\bar{\boldsymbol{\varepsilon}}^{HT} + \Delta\bar{\boldsymbol{\varepsilon}}^{V})
\tag{11.35}
$$

Fig. 11.2 Viscoelastic block
confined in a rigid die

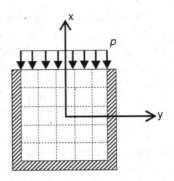

It is worth noticing that the viscoelastic strains at $t + \Delta t$ can be evaluated using the stresses obtained at the end of time t (see Sect. 3.2). So, the vector $\Delta \bar{\varepsilon}^V = \bar{\varepsilon}^V(t + \Delta t) - \bar{\varepsilon}^V(t)$ is known during the incremental step corresponding to the time interval $[t, t + \Delta t]$. Similarly, if we know the temperature and moisture histories, the incremental hygrothermal strain vector $\Delta \bar{\varepsilon}^{HT} = \bar{\varepsilon}^{HT}(t + \Delta t) - \bar{\varepsilon}^{HT}(t)$ is also known during that incremental step. Here, the overbar is being used to denote surface-averaged strain. The hygrothermal strain components at time t are given by

$$\varepsilon_i^{HT}(t) = \varepsilon_i^T(t) + \varepsilon_i^H(t) = \alpha_i \Delta \Theta(t) + \beta_i \Delta H(t) \qquad (11.36)$$

where β_i and $\Delta H(t)$ are, respectively, the hygroscopic expansion coefficient at i-direction and the moisture variation at time t in relation to the initial strain-free temperature. α_i and $\Delta \Theta(t)$ were defined in Sect. 6.1.

The global system of equations is obtained using the same procedures employed to find (11.24). Solving this global system, we determine the global incremental surface-averaged displacement vector $\Delta \bar{\mathbf{u}}_G$. The local incremental displacements, strains and stresses are obtained by the same procedures used for the case of linear elastic material. The updated displacements, strains and stresses of each subvolume are, then, given by

$$\mathbf{u}(t + \Delta t) = \mathbf{u}(t) + \Delta \mathbf{u} \qquad (11.37)$$

$$\varepsilon(t + \Delta t) = \varepsilon(t) + \Delta \varepsilon \qquad (11.38)$$

$$\sigma(t + \Delta t) = \sigma(t) + \Delta \sigma \qquad (11.39)$$

Example 1 To show an application of the *viscoelastic parametric finite-volume formulation*, we consider a viscoelastic block confined inside an infinitely rigid die and subjected to an instantaneously applied vertical pressure $p = 10\,MPa$ (Fig. 11.2). The material is isotropic, homogeneous, linear elastic in dilatation and linear viscoelastic in shear. This viscoelastic behavior is modeled by a Maxwell

Fig. 11.3 Horizontal normal stress in the viscoelastic block

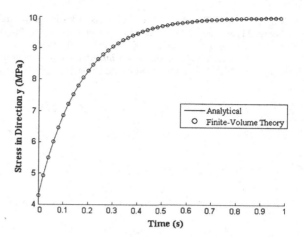

element with a spring of constant $G = 3,846.15\,MPa$ and a dashpot with $\eta_G = 400\,MPa.s$. The elastic bulk modulus is $K = 8,333.33\,MPa$. The friction effect along the interface block-die is neglected.

According to the analytical solution, the horizontal normal stress is given by [6]

$$\sigma_y(t) = -p\left\{1 - \frac{6G}{3K + 4G}\exp\left[-\frac{3Kt}{(3K + 4G)\theta}\right]\right\} \qquad (11.40)$$

where $\theta = \eta_G/G$. The block is under plane strain state.

To analyze the problem using the *viscoelastic parametric finite-volume formulation*, Escarpini Filho [5] discretized the block into rectangular subvolumes as shown in Fig. 11.2. The analytical and numerical results for the horizontal confinement stress in function of time are illustrated in Fig. 11.3. As observed, this parametric formulation provides results nearly identical to the exact analytical solution.

References

1. Y. Bansal, M.-J. Pindera, Efficient reformulation of the thermoelastic higher-order theory for FGMs. J. Therm. Stresses **26**(11/12), 1055–1092 (2003)
2. M.A.A. Cavalcante, S.P.C. Marques, M.-J. Pindera, Parametric formulation of the finite-volume theory for functionally graded materials-part I: analysis. J. Appl. Mech. **74**(5), 935–945 (2007)
3. M.A.A. Cavalcante, S.P.C. Marques, M.-J. Pindera, Parametric formulation of the finite-volume theory for functionally graded materials-part II: numerical results. J. Appl. Mech. **74**(5), 946–957 (2007)
4. M.A.A. Cavalcante, S.P.C. Marques, M.-J. Pindera, Computational aspects of the parametric finite-volume theory for functionally graded materials. Comput. Mater. Sci. **44**, 422–438 (2008)

5. R.S. Escarpini Filho, Analysis of linear viscoelastic composites structures using finite-volume theory. Master thesis. Federal University of Alagoas (in Portuguese) (2010)
6. W. Flügge, *Viscoelasticity* (Springer, New York, 1975)
7. H.K. Versteeg, W. Malalasekera, *An introduction to computational fluid dynamics—the finite volume method*, 2nd edn. (Pearson Education Limited, Harlow, 2007)

Chapter 12
Solutions with Abaqus

To help the reader to practice with a professional computer code, we use Abaqus to solve a few problems in viscoelasticity (small and large strains). First we relate Abaqus procedure to the general formulation given in this book and then we provide detailed instructions to run the code.

ABAQUS [1] is a highly sophisticated, general purpose finite element program, designed primarily to model the behavior of solids and structures under externally applied loading. It includes capabilities for geometrical modeling with an extensive element library for static and dynamic analyses in small and large deformation processes using linear and non linear constitutive relations. Its mechanical and computational theoretical background is up-to-date and well explained and it is widely used in industry and in academic research work.

12.1 Small Strain Examples

The formulation used in Abaqus for small strain problems is similar to the one in this book. Yet, some particularities are to be stressed.

From (4.3) and (4.7) in Chap. 4, for the relations in shear, we have (we are here using symbolic notation)

$$\mathbf{s} = \int_0^t 2\left(G_\infty + \sum_{i=1}^{n_G} G_i e^{-(t-\tau)/T_i} \right) \dot{\mathbf{e}} d\tau \tag{12.1}$$

Abaqus defines as state variables

$$\mathbf{q}_i(t) = \int_0^t \left(1 - e^{-(t-\tau)/T_i} \right) \dot{\mathbf{e}} d\tau \tag{12.2}$$

S. P. C. Marques and G. J. Creus, *Computational Viscoelasticity*,
SpringerBriefs in Computational Mechanics, DOI: 10.1007/978-3-642-25311-9_12,
© The Author(s) 2012

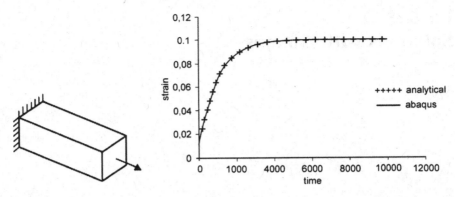

Fig. 12.1 Viscoelastic rod $30 \times 30 \times 100$ mm subjected to a creep test. Material elastic in bulk and viscoelastic in shear. Comparison of numerical and analytical results

and then, with $G_0 = G_\infty + \sum_{i=1}^{n_G} G_i$ and $g_i^P = G_i/G_0$, obtains

$$\mathbf{s} = 2G_0 \left(\mathbf{e} - \sum_{i=1}^{n} g_i^P \mathbf{q}_i \right) \qquad (12.3)$$

where q_i is determined with a recurrent algorithm as that in (3.17) assuming that $\dot{\mathbf{e}}$ in (12.2) varies linearly in each time interval. Similar procedure is used in relation to bulk.

To enter the material data into Abaqus, we have to inform elastic and viscoelastic properties. The elastic properties are given as the Young modulus E and the Poisson modulus v. To enter the viscoelastic properties we must give G_∞, G_i and T_i, $i = 1, \ldots, n_G$, being $G_\infty = G_0 - \sum_{i=1}^{n_G} G_i$; G_0 is determined using E and v and the expressions in Appendix B. The values of G_i are entered indirectly with the relation $g_i^P = G_i/G_0$; the values of T_i are entered directly. The same procedure is used for entering the bulk properties, where the viscoelastic parameter is k_i^P. Abaqus uses the same relaxation times for bulk and shear. The procedure will be detailed in Example 1 below.

Example 1 **Viscoelastic rod subjected to constant axial load** (Abaqus/Standard Example Problems Manual, Volume I).

This is a simple but interesting problem that introduces Abaqus viscoelastic procedures and explains how to find the parameters needed and to compare the answer with an analytical solution. We present it here with small modifications to adapt it to SI standards.

The rod has a length of 300 mm, and a square section with 30 mm in each side. A constant axial stress of 10 MPa is applied suddenly at one end and maintained constant, as shown in Fig. 12.1. We aim the history of deformations in time.

It is here assumed that the reader has a general knowledge of Abaqus/CAE (a tutorial *"Getting started with Abaqus interactive edition"* is available with the software). We will show here only the details related to the viscoelastic analysis.

Determination of material parameters: The material is represented by a Zener model with an **extensional** relaxation function

$$E(t) = k_1 + k_2 e^{-t/T^E} \tag{12.4}$$

Comparing this equation to (2.24) in Chap. 2, we see that $k_1 = E(\infty)$ and $k_2 = E(0) - E(\infty)$.

The relaxation time is $T^E = \eta/k_2$, where η is the viscosity. In this example $k_1 = 100\,MPa, k_2 = 900\,MPa$ and $T^E = 100\,s$. The bulk modulus $K = 10000\,MPa$ is independent of time.

Abaqus requires the elastic constants to be given in the *elastic* option through the elastic modulus E and the Poisson ratio v. The elastic modulus is immediately available as $E(0) = k_1 + k_2 = 1000 MPa$ and the Poisson ratio is determined as (from Appendix B)

$$v = \frac{3K - (k_1 + k_2)}{6K} = 0.4833$$

In the *viscoelastic* option, we have to use the relaxation functions for shear and bulk defined using Prony series. In this problem, no bulk relaxation occurs, but we need the shear relaxation function that Abaqus writes in the form

$$G(t) = G_0 \left(1 - \sum_{i=1}^{n^G} \bar{g}_i^P (1 - e^{-t/T_i^G}) \right) \tag{12.5}$$

The form used in this book is, for the Zener model, from (2.24) in Chap. 2,

$$G(t) = G_\infty + G_1 e^{-t/T_1^G} \tag{12.6}$$

Both equations are equivalent (check) because Abaqus defines $g_i^P = G_i/G_0$. Still, we have to obtain (12.5) from (12.4). In Example 3 from Chap. 5 the relation between extensional and shear relaxation functions was given. For the material characterized by (12.4) and purely elastic bulk behavior with bulk modulus K, we have from (4.20) or (5.37)

$$G(t) = \frac{27K^2 k_2}{(9K - k_1 - k_2)(9K - k_1)} \exp\left(-\frac{9K - k_1}{9K - k_1 - k_2} \frac{t}{T^E} \right) + \frac{3Kk_1}{9K - k_1} \tag{12.7}$$

Then, we have

$$G_0 = \frac{3K(k_1 + k_2)}{9K - k_1 - k_2} = 337.078; \quad G_\infty = \frac{3Kk_1}{9K - k_1} = 33.370; \quad G_1 = G_0 - G_\infty$$
$$= 303.707$$

and thus,

$$g_1^p = \frac{G_1}{G_0} = 0.901$$

$$T_1^G = \frac{9K - k_1 - k_2}{9K - k_1} T^E = 98.99$$

Entering data into Abaqus: we assume that the reader has a basic familiarity with Abaqus for the analysis of elastic problems. Thus, we include here only the details related to viscoelastic analyses. The example is called BeamV.

The characteristics of the **material** are given in module **Property.**

1. Entering the module, we go to **Create material. The Edit Material** dialog box appears.
2. Name the material Polymer.
3. From the **Material Behavior** menu select **Mechanical-Elasticity-Elastic.** Abaqus displays the **Elastic Data.** Maintain **Isotropic.**
4. **In Moduli Timescale (for viscoelasticity)** choose **instantaneous.**
5. Type the value 1000 for Young modulus and the value 0.4833 for Poisson ratio in the respective fields.
6. Without leaving this box, go again to **Mechanical, elasticity, viscoelastic.** In **Domain** choose **time,** in **time Prony.** Inform **gi:0.901, ki:0, tau:98.99.** If we have more terms in the Prony series we enter them below.
7. Save.

Defining and assigning section properties: Still in Module **Property,** select the tool **Create Section** to define a Bar section:

1. The **Create Section** dialog box appears.
2. In the **Create Section** dialog box:
2.1. Name the section BarVSection..
2.2. Accept **Category: Solid.**
2.3. **Type: Homogeneous.**
2.4. Click **Continue.** The **Edit Section** dialog box appears.
3. In the **Edit Section** dialog box:
3.1. In **Material** accept Polymer. If you had defined other materials, you could click the arrow next to the Material text box to see a list of available materials and to select the material of your choice.
3.2. Click **OK.**

Go to **Assign Section,** to assign the section BarSection, to the model.

1. Click with the mouse on the beam and press **DONE** under the viewport.
2. In the **Section Assignement** box verify **Section:** BeamVSection.
3. Click **OK.** The bar changes color.

Defining the assembly: This module may be used to assemble a complex model from its parts. In the present case, we have only one part. In the Module list located below the toolbar, click **Assembly** to enter the **Assembly module**.

1. From the menu bar, select **Instance Part**. The **Create Instance** dialog box appears.
2. In the **Instance Type** dialog box, choose **Independent** (mesh on instance).
3. In the dialog box, select **BeamV**, click OK.

Configuring the Analysis: Now that the assembly has been created, you can move to the **Step** module to configure your analysis. In the Module list, click **Step** to enter the **Step module**.
Select **Create Step.**

1. The **Create Step** dialog box appears with a list of all the general procedures and a step name. Write name **stepvisco**
2. Select **Procedure type:** General/Visco and click **Continue.**
3. The **Edit Step** Box opens. In the **Basic** tab, select **time period**: 10000, **NLGeom**-Off.
4. In the **Incrementation** tab, **Maximum number**: 1000; **increment size**: initial: 10, minimum: 0.001, maximum: 30. In the option **creep/swelling/viscoelastic strain error tolerance**: 0.005.
5. Click **OK** to create the step and to exit the **Edit Step** dialog box.

Then, we have to set the **boundary conditions encastré** in one of the ends and apply a **pressure -10** to the other, **mesh** the bar, choose the **element type** and **solve**.

Comparing the result to the analytical solution: In **Example 3** from Chap. 2, the creep function corresponding to a given relaxation function was determined. Thus, for the function (12.4) we obtain

$$D(t) = \frac{1}{k_1} - \frac{k_2}{k_1(k_1 + k_2)} e^{-t/\theta^E} \tag{12.8}$$

with

$$\theta^E = (1 + k_2/k_1)T^E = 1000$$

Abaqus gives a very approximate answer, as shown in Fig. 12.1 above.

Example 2 Change the integration parameters in Example 1 to see their influence on the approximation.

Example 3 Apply a constant displacement of 5 mm to the free end of the bar and determine the relaxation curve. Check the result with (12.4).

Example 4 Thick cylindrical shell under internal pressure-Plane strain.

This problem was solved using Laplace transform in **Example 5** of Chap. 5.

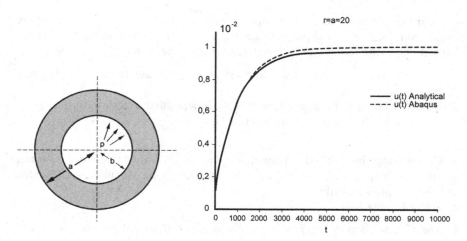

Fig. 12.2 Cylindrical tube subjected to internal pressure in plane strain. Dimensions: a = 20 mm, b = 10 mm. Pressure p = 10 MPa. Radial displacement of the external face

Dimensions: $b = 10$ mm, $a = 20$ mm; Material properties are the same as in Example 1: $E = 1000\,MPa$, $v = 0.483$, $g_i = 0.90101$ $k_i = 0$, $\tau_i = 98.99$ sec Pressure p = 10 MPa; Time step = 10000, increments: initial = 10, minimum = 0.1, maximum = 70, cetol = 0.005; Mesh size : 0.3, Element CPE4R.

Displacements are plotted in Fig. 12.2 compared to the analytical results given in Example 5 of Chap. 5. Stresses remain constant, as predicted analytically.

12.2 Thermo-Viscoelasticity Examples

Example 5 **Creep and relaxation tests at different temperatures**

We will analyze the same problem of Example 1 determining the changes due to increase of temperature. In this case we need to solve two problems: a thermal one, that provides the temperature distribution and a viscoelastic one, that uses the information of the first one to modify the constitutive parameters.

We begin the first problem **applying temperature** using the geometry of the beam.

In **Property** we inform **density:1, conductivity:1** and **specific heat:1**. In module **Step** go to **Create step,** select **heat transfer, Max. allowable temperature change per increment:** 0.0001. In module **load** go to **Create Load,** type at **Selected Step** temperature, select the beam and inform magnitude of temperature and **instantaneous.** In module **Mesh, Element Type** select **Heat Transfer.** In **Job** give the name **applyingtemperature.** Save as **applyingtemperature.**

Now we solve the viscoelatic problem, making the following changes (as related to Example 1):

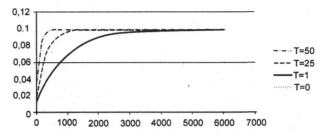

Fig. 12.3 Creep curves at different temperatures. Strain versus time in seconds

In **module Property**, after entering the viscoelastic data as before, we go to **Suboptions** choose **trs** (for temperature shift) and inform **theta0:0** (reference temperature), **C1:4.92 e C2:215**. In module **Load**, we keep the same boundary and load data, adding information from the. **odb** created previously. Still in the module **load** go to **Predefined Field in the upper** toolbar and choose **Create**. In **Category**, use **Other**, and in **Types for Selected Step** select **Temperature**, **continue**, select all the beam. In the box **Edit Predefined Field** select as distribution option **From results of output database file** and in file look for **applyingtemperature.odb** and press OK.

Then, continue as in the previous examples.

We observe in Figs. 12.3 and 12.4 that, as said in Chap. 6, the time shifting procedure does not alter the short term and long term behaviors but changes the retardation and/or relaxation times.

12.3 Finite Strain Examples

For large strain situations, Abaqus uses a procedure based in relation (8.12) (see Ciambella et al. [2, 3]. For the elastic deformation the hyperelastic equations are now used. The viscoelastic representation is the same as in the small strain part.

In Miller [4], an interesting application of Abaqus to Biomechanics may be found.

Example 6 **Uniaxial creep of a bar**

The same bar of Fig. 12.1 is now subjected to an axial stress of 50 MPa. Material behavior: in **Mechanical-elasticity-hyperelastic**. **Strain energy**: choose: Polynomial. **Input source**: choose: coefficients. **Strain energy potential order**: choose: 1; In **data** write: $C_{10} = 100$; $C_{01} = 68.5$; $D_1 = 0.0002$. Go to **Mechanical-elasticity-viscoelastic-** In **Domain** choose time, in **time** Prony. Inform: $gi = 0.901$, $ki = 0$, $tau = 98.99$. In module **step**, create step, choose **category** visco with **time period**:10000, **increment size**: initial: 0.1; minimum:0.1; maximum:50; **creep/swelling/viscoelastic strain error tolerance**: 0.1. In **other**, choose in **matrix storage** the option unsymmetric, keeping the other default options.

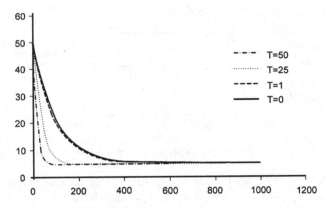

Fig. 12.4 Relaxation curves at different temperatures. Stress in MPa versus time in seconds

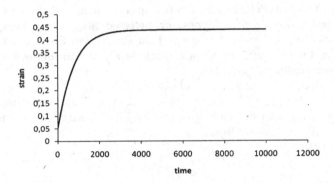

Fig. 12.5 Large visco-hyperelastic deformations for the rod under tension. Logarithmic strain versus time in seconds

The result, as a plot of logarithmic strain ($Log_{10}(1 + \varepsilon)$) versus time, is given in Fig. 12.5 above.

Situations with large strain are complex. In this example, if the load is too large the solution may become unstable. The use of the hybrid element C3D8RH may be convenient.

Example 7 **Punching of a viscoelastic foam.** This is an interesting example that is described in the Abaqus manual.

References

1. ABAQUS Theory Manual. Information on the SIMULIA site: http://www.simulia.com/support/documentation.html
2. J. Ciambella, M. Destrade, R. W. Ogden, On the ABAQUS FEA model of finite viscoelasticity. Rubber Chem. Technol. **82**(2), 184–193 (2009)

3. J. Ciambella, A. Paolone, S. Vidoli, A comparison of nonlinear integral-based viscoelastic models through compression tests on filled rubber. Mech. Mater. **42**, 932–944 (2010)
4. K. Miller, Constitutive model of brain tissue suitable for finite element analysis of surgical procedures. J. Biomech. **32**, 531–537 (1999)

Appendix A
Mathematical Formulae

In this appendix we present a brief review of some mathematical definitions, properties and relations that are useful for the development of several topics presented in this book. For a more detailed review, the reader is encouraged to consult specialized books (e.g., [1–3]).

1. Step and Impulse Functions
The unit step function $H(t)$ is defined as

$$H(t - t_0) = \begin{cases} 0 \text{ for } t < t_0 \\ 1 \text{ for } t > t_0 \end{cases} \tag{A.1}$$

The unit impulse or Dirac-delta function $\delta(t)$ has the following definition

$$\delta(t - t_0) = \begin{cases} 0 \text{ for } t \neq t_0 \\ \infty \text{ for } t = t_0 \end{cases} \tag{A.2}$$

with the additional condition $\int_{-\infty}^{\infty} \delta(t)dt = 1$.

2. Leibnitz Rule for the Differentiation of Integrals
Given

$$F(t) = \int_{g(t)}^{h(t)} f(t, \tau)d\tau \tag{A.3}$$

then

$$\frac{dF}{dt} = \int_{g(t)}^{h(t)} \frac{\partial f}{\partial t} d\tau - f(t, g)\frac{dg}{dt} + f(t, h)\frac{dh}{dt} \tag{A.4}$$

S. P. C. Marques and G. J. Creus, *Computational Viscoelasticity*,
SpringerBriefs in Computational Mechanics, DOI: 10.1007/978-3-642-25311-9,
© The Author(s) 2012

Table A.1 Laplace transform pairs

$f(t)$	$H(t)$	$\delta(t)$	e^{-at}	$\frac{1}{a}(1-e^{-at})$	$t^n(n=0,1...)$	$\sin at$	$\cos at$
$\bar{f}(s)$	$\frac{1}{s}$	1	1	$\frac{1}{s(a+s)}$	$\frac{n!}{s^{n+1}}$	$\frac{a}{s^2+a^2}$	$\frac{s}{s^2+a^2}$

When $h(t)=t$ and $g(t)=\tau_0=const$, (A.4) becomes

$$\frac{dF}{dt}=\int_{\tau_0}^{t}\frac{\partial f}{\partial t}d\tau+f(t,t) \tag{A.5}$$

3. Solution of the First Order Linear Differential Equation

The first order linear differential equation

$$\dot{y}+p(\tau)y=r(\tau) \tag{A.6}$$

with the initial condition $y(\tau_0)=y_0$ has the general solution

$$y(t)=\exp\left(-\int_{\tau_0}^{t}p(\tau)d\tau\right)\int_{\tau_0}^{t}\left[r(\tau)\exp\left(\int_{\tau_0}^{t}p(\tau)d\tau\right)\right]d\tau$$
$$+y_0\exp\left(-\int_{\tau_0}^{t}p(\tau)d\tau\right) \tag{A.7}$$

When $p(\tau)=p_0=const$, (A.7) becomes

$$y(t)=\exp[p_0(\tau_0-t)]\int_{\tau_0}^{t}r(\tau)\exp[p_0(\tau_0-t)]d\tau+y_0\exp[p_0(\tau_0-t)] \tag{A.8}$$

For $p(\tau)=p_0=const$ and $r(\tau)=r_0=const$, (A.8) yields

$$y(t)=\frac{r_0}{p_0}[1-\exp(p_0\tau_0-p_0t)]+y_0\exp(p_0\tau_0-p_0t) \tag{A.9}$$

4. Laplace Transform: Definition and Properties

Let $f(t)$ be a function of a real variable $t\geq0$. The Laplace transform of $f(t)$ is defined by

$$L\{f(t)\}=\bar{f}(s)=\int_{0}^{\infty}e^{-st}f(t)dt \tag{A.10}$$

where s is the transform parameter which may be complex or real.

Some Important Properties of Laplace Transform

(a) Linearity

For n functions $f_i(t)$, using the definition (A.10), it can be easily shown that

$$L\left\{\sum_{i=1}^{n} \alpha_i f_i(t)\right\} = \sum_{i=1}^{n} \alpha_i L\{f_i(t)\} \tag{A.11}$$

where α_i are values independent of t.

(b) Laplace transform of derivatives

If $f'(t)$ is the first derivative of $f(t)$, it is possible to prove that

$$L\{f'(t)\} = \bar{f}'(s) = s\bar{f}(s) - f(0) \tag{A.12}$$

For the case of the second derivative $f''(t)$, a similar procedure leads to the following equation

$$L\{f''(t)\} = \bar{f}''(s) = s^2\bar{f}(s) - sf(0) - f'(0) \tag{A.13}$$

Generally, for the derivative of order n, the following equation can be found

$$L\left\{f^{(n)}(t)\right\} = \bar{f}^{(n)}(s) = s^n\bar{f}(s) - s^{n-1}f(0) - s^{n-2}f'(0) \cdots - f^{(n-1)}(0) \tag{A.14}$$

(c) Laplace transform of integrals

The Laplace transform of the integral $g(t) = \int_0^t f(\tau)d\tau$ is given by

$$L\{g(t)\} = \bar{g}(s) = \int_0^\infty e^{-st} \int_0^t f(\tau)d\tau dt = \frac{\bar{f}(s)}{s} \tag{A.15}$$

(d) Laplace transform of the convolution of two functions

The convolution of two functions $f(t)$ and $g(t)$ defined for $t \geq 0$ is a new function $h(t)$ given by

$$h(t) = f(t) \circ g(t) = \int_0^t f(t)g(t - \tau)d\tau \quad (t \geq 0) \tag{A.16}$$

It can be easily shown, using a change of variable, that $f(t) \circ g(t) = g(t) \circ f(t)$. The Laplace transform of $h(t)$ is written as

$$L\{h(t)\} = \bar{h}(s) = \int_0^\infty e^{-st} \int_0^t f(t)g(t - \tau)d\tau dt = \bar{f}(s).\bar{g}(s) \tag{A.17}$$

This result is the well-known Convolution theorem.

(e) Laplace transforms of some common functions

Laplace transform pairs for some simple functions that often appear in applications are shown in Table A.1.

(f) Limit theorems

$$\lim_{s \to 0^+} s\bar{f}(s) = f(\infty) \tag{A.18}$$

$$\lim_{s \to \infty} s\bar{f}(s) = f(0) \tag{A.19}$$

Appendix B

For the sake of completeness and to establish a reference notation, a brief review of some continuum mechanics relations is included. For a complete formulation, books on linear and non linear continuum mechanics should be consulted (e.g., [4–6]).

B.1 Small Strain theory

The linear elastic constitutive relation is $\sigma = \mathbf{E}\varepsilon$. In *isotropic* small strain elasticity we may divide strains into *spherical* (or *volumetric*) and *deviatoric* (or *isochoric*) components

$$\varepsilon_0 = \frac{1}{3} tr\varepsilon \mathbf{I}$$

$$e = \varepsilon - \varepsilon_0 \tag{B.1}$$

Then, we can use (as an alternative to the Hooke law in terms of the elasticity modulus E and the Poisson modulus v) the relation

$$\sigma = 3\varepsilon_0 + 2G e \quad \text{or} \quad \sigma_{ij} = 3K\varepsilon_{ii}\delta_{ij} + 2Ge_{ij} \tag{B.2}$$

where

$$\sigma_0 = \frac{1}{3} tr\sigma \mathbf{I}$$

$$s = \sigma - \sigma_0 \tag{B.3}$$

are the *hydrostatic* and *deviatoric* stresses and K and G are the *bulk* and *shear* moduli respectively.

S. P. C. Marques and G. J. Creus, *Computational Viscoelasticity*,
SpringerBriefs in Computational Mechanics, DOI: 10.1007/978-3-642-25311-9,
© The Author(s) 2012

B.2 Relationship Among Elastic Constants

$$
\begin{aligned}
G &= \frac{E}{2(1+v)} = \frac{3K(1-2v)}{2(1+v)} = \frac{3KE}{(9K-E)} \\
v &= \frac{E}{2G} - 1 = \frac{(3K-2G)}{2(3K+G)} = \frac{(3K-E)}{6K} \\
E &= 2G(1+v) = \frac{9KG}{(3K+G)} = 3K(1-2v) \\
K &= \frac{2G(1+v)}{3(1-2v)} = \frac{GE}{3(3G-E)} = \frac{E}{3(1-2v)}
\end{aligned}
\tag{B.4}
$$

B.3 Finite Deformations

$\mathbf{X}$ and $\mathbf{x}$ represent, respectively, the initial and actual positions of a material point. The deformation of the solid is represented by a mapping $\mathbf{x}(t) = \hat{\mathbf{x}}(\mathbf{X}, t)$.

Locally the mapping is approximated by the *deformation gradient* $\mathbf{F}$

$$
\mathbf{F}(\mathbf{X}, t) = \frac{\partial \hat{\mathbf{x}}}{\partial \mathbf{X}}
\tag{B.5}
$$

that is the basic measure in the large strain theory. The *displacement* is given by

$$
\mathbf{u}(\mathbf{X}, t) = \mathbf{x} - \mathbf{X}
\tag{B.6}
$$

and thus, the *gradient of displacement* is

$$
\frac{\partial \mathbf{u}}{\partial \mathbf{X}} = \mathbf{F} - \mathbf{I}
\tag{B.7}
$$

where $\mathbf{I}$ is the *unit tensor*.

$J = \det \mathbf{F}$ is the *volume ratio* or *Jacobian determinant*.

The deformation gradient tensor may be split into pure deformation and pure rotation components

$$
\mathbf{F} = \mathbf{RU} = \mathbf{VR}
\tag{B.8}
$$

with $\mathbf{R}^T = \mathbf{R}^{-1}$, $\mathbf{U} = \mathbf{U}^T$ and $\mathbf{V} = \mathbf{V}^T$.

$$
\mathbf{C} = \mathbf{F}^T \mathbf{F} = \mathbf{U}^T \mathbf{U}
\tag{B.9}
$$

is the *right Cauchy-Green tensor* and

$$
\mathbf{E} = \frac{1}{2}(\mathbf{C} - \mathbf{I}) \text{ or } E_{ij} = \frac{1}{2}\left(\frac{\partial u_i}{\partial X_j} + \frac{\partial u_j}{\partial X_i} + \frac{\partial u_k}{\partial X_i}\frac{\partial u_k}{\partial X_j}\right)
\tag{B.10}
$$

is the *Green-Lagrange strain tensor*. In the small strain case we have $\dfrac{\partial u_k}{\partial X_i} \ll 1$. and we write

$$E_{ij} \cong \varepsilon_{ij} = \frac{1}{2}\left(\frac{\partial u_i}{\partial X_j} + \frac{\partial u_j}{\partial X_i}\right) \tag{B.11}$$

B.4 Stress Measures

σ measured on the deformed configuration is the *true stress* or *Cauchy stress*. Other stress measures, useful in large strain calculations, are

$$\tau = J\sigma : Kirchhoff \text{ stress} \tag{B.12}$$

$$\mathbf{P} = J\sigma\mathbf{F}^{-T} : (\text{non-symmetric})First\,Piola - Kirchhoff \text{ stress or nominal stress} \tag{B.13}$$

$$\mathbf{S} = \mathbf{F}^{-1}\mathbf{P} = \mathbf{F}^{-1}\tau\mathbf{F} = J\mathbf{F}^{-1}\sigma\mathbf{F}^{-T} : Second\,Piola - Kirchhoff \text{ stress} \tag{B.14}$$

In the small strain case we have $\mathbf{F} \cong \mathbf{I}$, $J \cong 1$ and all stress measures coincide.

Appendix C

Here, we show some matrices appearing in the Parametric Finite-Volume Formulation presented in Chap. 11. The matrices used to define $\bar{\mathbf{A}}$ in Eq. (11.11) are given by

$$\mathbf{A} = \begin{bmatrix} \mathbf{A}^{(1)} \\ \mathbf{A}^{(2)} \\ \mathbf{A}^{(3)} \\ \mathbf{A}^{(4)} \end{bmatrix} \quad \mathbf{B} = \begin{bmatrix} \hat{\mathbf{J}} & 0 & 0 & 0 & 0 & 0 & 0 & 0 \\ 0 & \hat{\mathbf{J}} & 0 & 0 & 0 & 0 & 0 & 0 \\ 0 & 0 & \hat{\mathbf{J}} & 0 & 0 & 0 & 0 & 0 \\ 0 & 0 & 0 & \hat{\mathbf{J}} & 0 & 0 & 0 & 0 \\ 0 & 0 & 0 & 0 & \hat{\mathbf{J}} & 0 & 0 & 0 \\ 0 & 0 & 0 & 0 & 0 & \hat{\mathbf{J}} & 0 & 0 \\ 0 & 0 & 0 & 0 & 0 & 0 & \hat{\mathbf{J}} & 0 \\ 0 & 0 & 0 & 0 & 0 & 0 & 0 & \hat{\mathbf{J}} \end{bmatrix} \quad (C.1)$$

where

$$\mathbf{A}^{(1,3)} = \begin{bmatrix} 1 & 0 & 0 & 0 & 0 & 0 & 0 & 0 \\ 0 & 1 & 0 & \mp 3 & 0 & 0 & 0 & 0 \\ 0 & 0 & 0 & 0 & 1 & 0 & 0 & 0 \\ 0 & 0 & 0 & 0 & 0 & 1 & 0 & \mp 3 \end{bmatrix}$$

$$\mathbf{A}^{(2,4)} = \begin{bmatrix} 1 & 0 & \pm 3 & 0 & 0 & 0 & 0 & 0 \\ 0 & 1 & 0 & 0 & 0 & 0 & 0 & 0 \\ 0 & 0 & 0 & 0 & 1 & 0 & \pm 3 & 0 \\ 0 & 0 & 0 & 0 & 0 & 1 & 0 & 0 \end{bmatrix}$$

$$\mathbf{H} = \begin{bmatrix} \hat{\mathbf{C}} & 0 & 0 & 0 \\ 0 & \hat{\mathbf{C}} & 0 & 0 \\ 0 & 0 & \hat{\mathbf{C}} & 0 \\ 0 & 0 & 0 & \hat{\mathbf{C}} \end{bmatrix} \quad \mathbf{D} = \begin{bmatrix} \mathbf{n}^{(1)} & 0 & 0 & 0 \\ 0 & \mathbf{n}^{(2)} & 0 & 0 \\ 0 & 0 & \mathbf{n}^{(3)} & 0 \\ 0 & 0 & 0 & \mathbf{n}^{(4)} \end{bmatrix} \quad (C.3)$$

S. P. C. Marques and G. J. Creus, *Computational Viscoelasticity*, SpringerBriefs in Computational Mechanics, DOI: 10.1007/978-3-642-25311-9, © The Author(s) 2012

being

$$\mathbf{n}^{(k)} = \begin{bmatrix} n_1^{(k)} & 0 & n_2^{(k)} \\ 0 & n_2^{(k)} & n_1^{(k)} \end{bmatrix} \tag{C.4}$$

$$\mathbf{L} = \begin{bmatrix} \bar{L} & 0 & 0 & 0 \\ 0 & \bar{L} & 0 & 0 \\ 0 & 0 & \bar{L} & 0 \\ 0 & 0 & 0 & \bar{L} \end{bmatrix} \tag{C.5}$$

The matrices $\mathbf{P}$, $\mathbf{M}$ and $\mathbf{N}$ that appear in (11.19) are defined by

$$\mathbf{P} = \begin{bmatrix} 0 & 0 & 1/2 & 0 & 0 & 0 & -1/2 & 0 \\ -1/2 & 0 & 0 & 0 & 1/2 & 0 & 0 & 0 \\ 0 & 0 & 1/2 & 1/2 & 0 & 0 & 1/2 & 0 \\ 1/2 & 0 & 0 & 0 & 1/2 & 0 & 0 & 0 \\ 0 & 0 & 0 & 1/2 & 0 & 0 & 0 & -1/2 \\ 0 & -1/2 & 0 & 0 & 0 & 1/2 & 0 & 0 \\ 0 & 0 & 0 & 1/2 & 0 & 0 & 0 & 1/2 \\ 0 & 1/2 & 0 & 0 & 0 & 1/2 & 0 & 0 \end{bmatrix} \tag{C.6}$$

$$\mathbf{M} = \begin{bmatrix} 0 & 0 & 1 & 0 & 0 & 0 & 1 & 0 \\ 1 & 0 & 0 & 0 & 1 & 0 & 0 & 0 \\ 0 & 0 & 0 & 1 & 0 & 0 & 0 & 1 \\ 0 & 1 & 0 & 0 & 0 & 1 & 0 & 0 \end{bmatrix} \quad \mathbf{N} = \begin{bmatrix} 0 & 0 \\ 0 & 0 \\ 1 & 0 \\ 1 & 0 \\ 0 & 0 \\ 0 & 0 \\ 0 & 1 \\ 0 & 1 \end{bmatrix} \tag{C.7}$$

References

1. G.B. Arfken, H.J. Weber, *Mathematical Methods for Physicists* (Academic Press, San Diego, 2001)
2. A.D. Myskis, *Advanced Mathematics for Engineers* (Mir Publishers, Moscow, 1979)
3. C.R. Wylie, L.C. Barrett, *Advanced Engineering Mathematics* (McGraw-Hill, New York, 1995)
4. G.A. Holtzapfel, *Nonlinear Solid Mechanics* (Wiley, West Sussex, 2004)
5. L.E. Malvern, *Introduction to the Mechanics of Continuous Media* (Prentice-Hall, Englewood Cliffs, 1969)
6. C. Truesdell, W. Noll, *The Non-Linear Field Theories of Mechanics* (Springer, Berlin, 2004)

Index

S. P. C. Marques and G. J. Creus, *Computational Viscoelasticity*,
SpringerBriefs in Computational Mechanics, DOI: 10.1007/978-3-642-25311-9,
© The Author(s) 2012